Maturation of Neurotransmission

Satellite Symposium to the Sixth Meeting of the International Society
for Neurochemistry on Maturational Aspects of Neurotransmission Mechanisms,
Saint-Vincent, Aosta, August 29–31, 1977

Maturation of Neurotransmission

Biochemical Aspects

Editors
A. Vernadakis, Denver, Colo., *E. Giacobini*, Storrs, Conn., and
G. Filogamo, Turin

90 figures and 11 tables, 1978

S. Karger · Basel · München · Paris · London · New York · Sydney

Cataloging in Publication
   Maturation of neurotransmission: biochemical aspects
   Editors: A. Vernadakis, E. Giacobini, and G. Filogamo
   Basel; New York: Karger, 1978
   Satellite symposium to the sixth meeting of the International Society for Neurochemistry on maturational aspects of neurotransmission mechanisms, Saint-Vincent, Aosta, August 29–31, 1977
   1. Nervous System – growth and development – congresses 2. Neural Transmission – congresses 3. Neurochemistry – congresses
   I. Vernadakis, Antonia, 1930– ed. II. Giacobini, Ezio, ed. III. Filogamo, Guido, ed. IV. International Society for Neurochemistry V. Title: Satellite symposium on maturation of neurotransmission VI. Title
   WL 104 M445 1977
   ISBN 3–8055–2833–7

All rights reserved.
No part of this publication may be translated into other languages, reproduced or utilized in any form or by any means, electronic or mechanical, including photocopying, recording, microcopying, or by any information storage and retrieval system, without permission in writing from the publisher.

© Copyright 1978 by S. Karger AG, 4011 Basel (Switzerland), Arnold-Böcklin-Strasse 25
Printed in Switzerland by Thür AG Offsetdruck, Pratteln
ISBN 3–8055–2833–7

Contents

Foreword VII

Maturation of Cellular and Molecular Mechanisms

Filogamo, G.; Peirone, S., and Sisto Daneo, L. (Turin): How Early Do Myoblast Determination and Fast and Slow Muscle Fiber Differentiation Occur? 1

Gombos, G.; Ghandour, M.S.; Vincendon, G.; Reeber, A., and Zanetta, J.-P. (Strasbourg): Formation of the Neuronal Circuitry of Rat Cerebellum: Plasma Membrane Modifications 10

Giacobini, G. (Turin): Neuromuscular Alterations as a Consequence of Cholinergic Blockade during Development 23

Gardner, J.M. and Fambrough, D.M. (Baltimore, Md.): Properties of Acetylcholine Receptor Turnover in Cultured Embryonic Muscle Cells 31

Biosynthesis of Neurotransmitters and Enzymatic Regulation

Giacobini, E. (Storrs, Conn.): Regulation of Neurotransmitter Biosynthesis during Development in the Peripheral Nervous System 41

Black, I.B. and Coughlin, M.D. (New York, N.Y.): Ontogeny of an Embryonic Mouse Sympathetic Ganglion *in vivo* and *in vitro* 65

Otten, U.; Goedert, M., and Thoenen, H. (Basel): Role of Nerve Growth Factor for the Development and Maintenance of Function of Sympathetic Neurons and Adrenal Medullary Cells 76

Ossola, L.; Maitre, M.; Blindermann, J.M., and Mandel, P. (Strasbourg): Some Aspects of GABA Level Regulation in Developing Rat Brain 83

Koenig, J. and Koenig, H.L. (Paris): One Molecular Form of AChE Associated with Synapses in Two Cholinergic Systems: Skeletal Muscle of the Rat and Ciliary Ganglion of the Chick 91

Suzuki, O. and Yagi, K. (Nagoya): Multiple Forms of Monoamine Oxidase in the Human Cerebral Cortices at Different Ages 100

Uptake, Storage and Transport of Neurotransmitters

Hösli, E. and Hösli, L. (Basel): Uptake of ^{3}H-GABA in Organotypic Cultures of Fetal Human and Newborn Rat Nervous Tissue 108
Bondy, S.C.; Harrington, M.E.; Nidess, R., and Purdy, J.L. (Denver, Colo.): Neurotransmission-Related Development of the Chick Optic Lobe: Effects of Denervation, Noninnervation and Sensory Deprivation 116
Marchisio, P.C. and Di Renzo, M.F. (Turin): Axonal Transport during the Development of Retinotectal Connections in Chick Embryo 127
Coyle, J.T.; Campochiaro, P., and London, E.D. (Baltimore, Md.): Neurotoxic Action of Kainic Acid in the Developing Rat Striatum 134

Effects of Intrinsic and Extrinsic Factors in Neuronal Maturation

Pierce, T.; Hanson, R.K.; Deanin, G.G.; Gordon, M.W., and Levi, A. (Norwich, Conn.): Developmental and Biochemical Studies on Tubulin:Tyrosine Ligase 142
Festoff, B.; Duell, M.J., and Fernandez, H.L. (Kansas City, Mo.): Trophic Effects of Axonally Transported Proteins on Muscle Cells in Cultures 152
Vernadakis, A.; Arnold, E.B., and Hoffman, D.W. (Denver, Colo.): Neural Tissue Culture: A Model for the Study of the Maturation of Neurotransmission 160
Lauder, J.M. and Krebs, H. (Storrs, Conn.): Serotonin and Early Neurogenesis 171
Timiras, P.S. and Vaccari, A. (Berkeley, Calif.): Adaptive Changes Induced by Environmental and Hormonal Factors on the Development of Brain Neurotransmitter Systems 181
Tissari, A.H. and Tikkanen, I.T. (Helsinki): Maturation of the Responses of Brain 5-Hydroxytryptamine Turnover, Plasma Nonesterified Fatty Acids and Corticosterone to Stress during Ontogeny 191

Development of Biochemical Correlates of Behavior

Sparber, S.B. (Minneapolis, Minn.): Is Prenatal Induction of Tyrosine Hydroxylase Associated with Postnatal Behavioral Changes? 200
Lytle, L.D. and Meyer, E., jr. (Cambridge, Mass.): Developmental Neurochemical and Behavioral Effects of Drugs 210
Kellogg, C. (Rochester, N.Y.): Neurotransmitter Interactions and Early Convulsive Activity. A Developmental Model of Behavioral Responsivity 217
Lundborg, P. and Engel, J. (Göteborg): Neurochemical Brain Changes Associated with Behavioural Disturbances after Early Treatment with Psychotropic Drugs 226

Author Index 236

Foreword

The transfer of signals between individual cells during development is the subject of widespread research today. The papers collected in this volume focus primarily upon early interactions between nerve cells as well as interactions between nerve cells and their targets. In addition, maturational changes in the development of the nervous system, encompassing the period from neural tube induction until birth, are examined on both the cellular and organismic levels. Interferences in the normal progression of these maturational processes, such as those posed by hormonal, nutritional, and internal or external environmental influences, are also considered. The communications presented at this symposium have been prepared by experts in these fields, and, while they do not attempt to resolve all questions relating to the maturation of neurotransmissional mechanisms, they certainly succeed in exploring some of the more fundamental aspects of these processes.

This Symposium was held in the Center of Congress of St. Vincent (Valle d'Aosta) and was organized by *Antonia Vernadakis, Ezio Giacobini*, and *Guido Filogamo*. The secretarial assistance of Dr. *Giacomo Giacobini* was invaluable in the success of this Conference. We are indebted to several groups for their assistance in organizing this meeting. We wish to express our gratitude to a number of public and private organizations, especially the Società Incremento Turismo Valle d'Aosta organization in St. Vincent, for providing substantial financial support. We would also like to thank the authors for having accepted the additional effort of preparing the manuscripts contained in this volume; and last, but not least, we would like to thank S. Karger Publishers for having taken on and carried out the task of publishing this book. Finally, we are greatly indebted to Drs. *Ellena Camino* and *Maria G. Robecchi,* who helped in various ways to make this conference possible.

G. Filogamo

Maturation of Cellular and Molecular Mechanisms

Maturation of Neurotransmission. Satellite Symp., 6th Meeting Int. Soc. Neurochemistry, Saint-Vincent 1977, pp. 1–9 (Karger, Basel 1978)

How Early Do Myoblast Determination and Fast and Slow Muscle Fiber Differentiation Occur?[1]

G. Filogamo, S. Peirone and L. Sisto Daneo

Institute of Human Anatomy and Institute of Veterinary Anatomy, University of Turin, Turin

The dependence of muscle primordia differentiation upon signals emanated by growing axons seems to occur at several steps during epigenetic development, in which manifold interactions between neural units and their target structures take place. Due to limited space two critical points will be focused on here: (1) myoblast determination, and (2) the developing characteristic properties of slow and fast muscle fibres.

This presentation will consist of three parts: (1) chronology of the encounter *in vivo* between the somite mesodermal cells and growth cones of young neurons; (2) elucidation of the significance of this encounter by experiments *in vitro;* (3) fast and slow muscle fibre differentiation 'in myotome'.

(1) That myogenic line cells are related to young motor neurons 'very early' in the embryo is well established (5, 6), but 'how early' is not yet known. Answers to 'how early' (9) may help to elucidate whether the outgrowing axons from the neural tube govern the determination of myoblasts. This remains a puzzling question. In the chick embryo, the critical moment when the nerve processes make contact with the somites occurs at 35–40 h, 12th somite stage (stage 11 HH). Some cells of the lateral wall of the neural tube send out filopodia which, crossing the perineural space, contact the cells of the medial somite wall; these may likewise be provided with filopodia (fig. 1). At the level of the cranial somites, filopodia are numerous, while there are few at the level of the caudal somites.

In 53-hour embryos (stage 16 HH), many filopodia are still present; conversely, at the thoracolumbar level, a slender motor root has already formed. This consists of primitive axons devoid of microtubules and simply containing fine filaments and some vesicles.

[1] This work is dedicated to Prof. *Otto Bucher*. It was supported through a grant by the Consiglio Nazionale delle Ricerche No. 76.01398.04.

Two questions are raised at once by our observations: (1) which part of the somite is invested by neural projections and (2) what change, if any, is undergone by the components of this part? So far, from serially cut somites, we have found (9) that the somites at 35–40 h are divided into four sections of different cellular types: one characterizing the somite caudal half, two others the latero-dorsal and medial wall respectively, and one the somite core.

The cells of the medial somite wall, contacted by filopodia as early as the 12th somite stage, show different morphological features from the other three somite sections, two of which consist of prospective dermatome and sclerotome cells, respectively, while the differentiation of the somite core is further delayed (11). In the caudal somites at 35–40 h the above differences are barely perceptible. The tactical behaviour of the cells of the medial somite wall helps to indicate that they now can be determined as immature myoblasts, precursors of the mature ones which attain a higher degree of differentiation. As is known, somites are transient structures; when their 'breakage' occurs they become demarcated into a dermatome, a sclerotome and a myotome. The cells of the medial wall move outward, either individually or in groups, and layer up at the deep face of the dermatome to form the myotome. While still migrating, the immature myoblasts, mixed with undetermined mesodermal cells, are seen to be in contact with laterally growing axonal processes which extend out of the neural tube (fig. 2). These processes will graze, but not penetrate, the myotome. Coincident with the nerve fibres making contact with the myotome, the immature myoblasts appear clearly differentiated as mature myoblasts, fusing at once into multinucleated myotubes. This 'in myotome' maturation, which implies interactions between myoblasts, is accompanied by structural and functional changes as *in vitro* (2, 3, 10, 12, 15, 16); namely, the first appearance of the Koelle reaction for AChE (4) and ACh-R and ChA, the development of the cell membrane invaginations to form the T-systems, and the onset of contractile properties.

Further morphological observations, carried out in the formation centres of new myogenic primordia (limb buds) (8) at day 5 of incubation, have confirmed the exact chronological coincidence between nerve fibre growth cones making contact with mesodermal cells and the first signs of transformation of the latter into immature myoblasts, with the very earliest moments of their assemblage and orientation. One can now assume that the determination of immature myoblasts in the medial wall of the somite, and their subsequent migration into the myotome, involves the constructive influence of neural filopodia.

Fig. 1. 35-hour chick embryo; transverse section of the 9th somite. Some cells of the lateral wall of the neural tube (NT) sprout filopodia toward the medial wall of the somite and contact it. MC = Mesodermal cells. × 12,500.

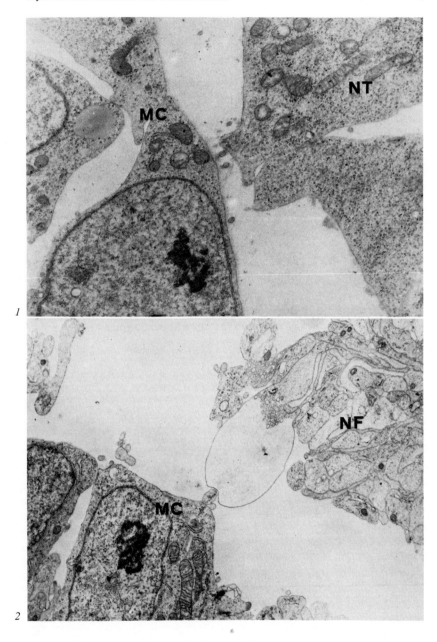

Fig. 2. 4-day chick embryo; transverse section at the level of thoracic somite. A mesodermal cell (MC) contacted by motor root fibre terminals (NF) elongating across the perineural space. × 7,400.

To investigate the first step of the epigenetic development of the myogenic line, we have undertaken the study of the *in vitro* evolution of explants of early somites, either alone or associated with neural tissue.

(2) The problem then arises in culturing somites as yet, *in vivo*, uncontacted by nerve fibres, since it is mandatory to avoid using immature myoblasts that are unquestionably 'automatic' (even in the absence of nerve fibres) precursors of mature myoblasts.

As is known, myoblast precursors obtained from 10- to 12-day-old chicken embryo muscles develop *in vitro* to mature myoblasts which, in the presence of calcium ions, fuse and form myotubes.

Experiments performed by *Sartore et al.* (14) have shown that before fusion, in immature myoblasts, a dramatic rearrangement of the cell surface structure takes place. This phenomenon is the counterpart, at the cell surface level, of the cytoplasmic differentiation that leads to the appearance of specific gene products, such as muscle type myosin. Under critical calcium ion concentrations myoblast fusion is blocked. However, both cell surface structure rearrangement and muscle type myosin synthesis do occur normally and lead to the formation of what we call a 'mature' myoblast, still a mononucleated cell, but with molecular features of fully differentiated muscle cells.

Accordingly, on the basis of the previous observations of *Ellison et al.* (1) and *Peirone et al.* (13), we set up two culture groups: in the first culture group we excised and cultured the most caudal of the ten somites present at stage 10, partly alone and partly associated with explants of the neural tube from the cranial region of the same embryos. In the second culture group all the somites from stage 9 embryos (9 somites) were excised, pooled together, and then divided into two subgroups: one of these was cultured alone, the other was grown together with neural tubes from 5-day embryos.

Evidence was gained that, by culturing both the most caudal somites from stage 10 embryos and all the somites at stage 9, the appearance of mononucleate elements displaying typical *in vitro* myoblast features was elicited only in the presence of axonal processes sprouting from neurons developing in the neural tube (fig. 3–4).

In the same cultures these myoblasts undergo maturation, then fuse giving rise to myotubes. Conversely, by culturing stage 9 to 10 somites associated with spinal ganglia from 5-day embryos, myoblasts and myotubes failed to differentiate, despite the growth of sensory nerve fibres among the mesodermal cells.

The above experiments strongly suggest that immature myoblast determination from mesodermal cells *in vitro* is triggered by some guiding agent(s) emanating from the growing neurons within the neural tube. The subsequent differentiation into mature myoblasts and myotubes is a wholly autonomous process, though liable to modulation by nerve fibres. The shift of cell surface

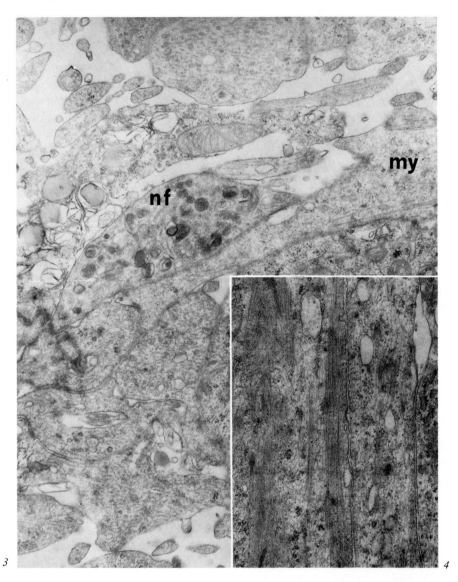

Fig. 3. 7-day culture of chick embryo somitic mesodermal cells associated to explants of neural tube showing a neuromuscular contact. my = Myoblast, nf = nervous fibre. × 30,000.

Fig. 4. 7-day culture of chick embryo somitic mesodermal cells associated to explants of neural tube showing two muscular elements. × 21,000.

binding properties toward the probe TNBS (2,4,6-trinitrobenzene sulphonic acid) is a very sensitive tool for monitoring muscle cell differentiation, since it detects a maturation event that occurs in the myoblast pathway of differentiation, before, and independently of, morphological changes of the cell structure.

By this technique we have recently shown that somite mesodermal cells, dissected before priming by exploring nerve fibres, do not differentiate into 'mature' myoblasts, even if grown for as long as 7 days *in vitro*. In identical culture conditions myoblast precursors, obtained from muscle primordia already primed by motor neuron exploring fibres, differentiate, showing the shift of TNBS binding properties peculiar to 'mature' myoblasts and myotubes (7).

(3) There is general (though not unanimous) agreement that fast and slow muscle fibres are not inherently different; their difference seems to be a response to the influence exerted by innervating motor axons (possibly by the amount of transmitter they release or by the release of some trophic substance) at the moment of the synaptic area differentiation. This area attains maturity following different maturation patterns in fast and slow fibres, respectively. A member of our group (17) has recently shown that, as early as day 6 of incubation, two different types of myoblasts and myotubes make up the primordia of the anterior latissimus dorsi (ALD) and the posterior latissimus dorsi (PLD) muscles, respectively. In these primordia, the relationship between nerve fibres and one or the other type of myoblast-myotubes is seen to differ. It is evident, in fact, that nerve fibres travel in close proximity with the slow-type myoblast-myotubes, and frequently end into them, whereas they keep at some distance from fast-type myoblast-myotubes and never make contact with them, but are clearly isolated from them, by Schwann cell and fibroblast lamellar expansions.

The stage at which fast and slow fibres are believed to differentiate (day 5–6), is actually a very early one, so it was quite unexpected to observe two morphologically identifiable myoblast and myotube types even earlier 'in myotome' (fig. 5), that is, on the 3rd day of incubation. Indeed, these two types are similar to those characterizing the ALD and PLD, respectively, on day 6.

In the myotome, the myoblasts similar to those of the fast type are located in a dorsal position, whereas those of the slow type lie in a ventral position; the latter, therefore, are consistently in close proximity to nerve fibres. This is one area of inquiry that, due to lack of space, we shall not develop any further here. The elementary point to be made here is that we have demonstrated 'how early' the young neurons contribute to some modulation of myoblast differentiation by virtue of their intrinsic guiding factors.

Fig. 5. 53-hour chick embryo. Two types of myoblasts are peculiar of anterior (a) and posterior (b) part of the myotome. They are similar to rapidly twitching and slowly twitching contractile elements. $a \times 8,700$. $b \times 8,000$.

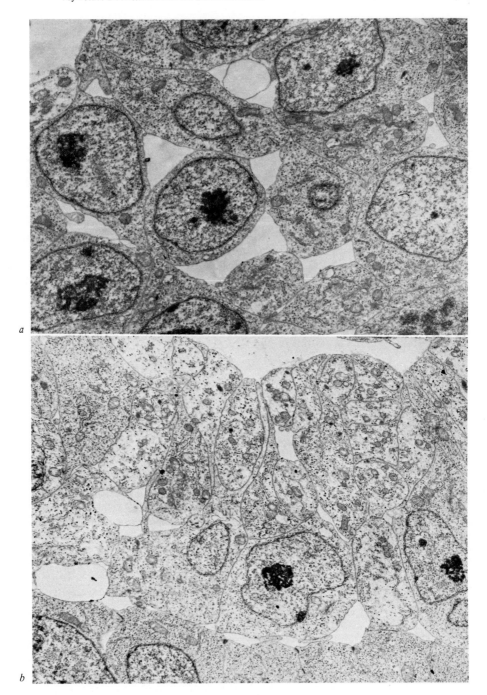

Conclusions

It is known that some embryological interactions at the neuromuscular level occur at increasingly earlier steps in development, concurrently with their being brought into focus. Until recently, most studies suggested that slow and fast fibres are neurally modulated in the synaptic period; a large body of evidence provided here indicates that in the chick embryo they are present in the myotome as early as the third day of incubation.

The more interesting outcome of our investigations is that young motor neurons may be responsible for the determination of immature myoblasts from mesodermal cells as early as stage 11 HH (12 somites). Nerve fibres, therefore, seem to exert an influence on the mesodermal somite cells in the selection of their evolutionary line.

Several questions arise regarding the mechanism of these precocious neural influences: (1) what is the relevance of the cholinergic system present as early as 40–60 h of incubation?; (2) are neurons, in the absence of organized synapses, capable of bioelectric discharge in the period of the first contacts and of global innervation?; (3) are the filopodia and growth cones of the young neurons the source of active molecules?

We believe that the field of precocious interactions has come to the stage in which more analytical work is required on the nature of some directive factors.

Summary

The determination of the myoblasts by nerve fibres and the influence of these fibres on muscle differentiation were investigated in chick embryos.

Using serially cut embryos examined by electron microscopy, evidence was gained that very early (35–40 h of incubation or 11 HH stage), medial somite cells are reached by nerve filopodia and axons growing from the neural tube. Furthermore, an exact coincidence between the reaching of the mesodermal somite cells by the axons and the appearance of the first myoblasts is brought to light. *In vitro* experiments performed by culturing somites, alone or associated with neural tube explants, have confirmed that the nerve fibres, coming into contact with mesodermal cells, seem to exert an influence on their subsequent myogenic evolution.

Regarding the influence of motor neurons on muscle differentiation, two different types of myotubes, resembling the fast and slow of the ADL and PDL primordia, have been found in the myotome on the third incubation day. The possible influence of the nerve fibres on the precocious differentiation of 'fast and slow' is also discussed.

References

1 *Ellison, M.L.; Ambrose, E.J., and Easty, G.C.:* Myogenesis in chick embryo somites *in vitro.* J. Embryol. exp. Morph. *21:* 341–346 (1969).

2   *Ezerman, E.B. and Ishikawa, H.:* Differentiation of the sarcoplasmic reticulum and T-system in developing chick skeletal muscle *in vitro.* J. Cell Biol. *35:* 405–420 (1967).
3   *Fambrough, D. and Rash, J.E.:* Development of acetylcholine sensitivity during myogenesis. Devl Biol. *26:* 55–68 (1971).
4   *Filogamo, G.:* Activité AChE au niveau des jonctions neuromusculaires. C.r. Ass. Anat., Bruxelles *1963:* 115–121.
5   *Filogamo, G. and Gabella, G.:* The development of neuro-muscular correlations in vertebrates. Archs Biol., Liège *78:* 9–60 (1967).
6   *Filogamo, G. and Marchisio, P.C.:* Acetyl-choline system and neural development. Neurosci. Res. *4:* 29–64 (1971).
7   *Filogamo, G.; Peirone, S., and Comoglio, P.M.:* In preparation (1977).
8   *Filogamo, G. and Sisto Daneo, L.:* Pioneering nerve fibres in the limb bud. 4th Congr. Eur. Anat., Basel 1977.
9   *Filogamo, G. and Sisto Daneo, L.:* Nervous projections and myotome maturation. J. submicrosc. Cytol. *9:* 307–310 (1977).
10 *Fischman, D.A.:* Development of striated muscle; in *Bourne* The structure and function of muscle; 2nd ed., vol. 1, pp. 75–148 (Academic Press, New York 1972).
11 *Panattoni, G. and Sisto Daneo, L.:* The tridimensional model of chick embryo somites. 4th Congr. Eur. Anat., Basel 1977.
12 *Paterson, B. and Prives, J.:* Appearance of acetylcholine receptors in differentiating cultures of embryonic chick breast muscle. J. Cell Biol. *59:* 2241–2245 (1973).
13 *Peirone, S.; Sisto Daneo, L., and Filogamo, G.:* Myogenic imprinting of early somites by nerve fibres *in vitro.* J. submicrosc. Cytol. *9:* 311–314 (1977).
14 *Sartore, S.; Tarone, G.; Cantini, M.; Schiaffino, S., and Comoglio, P.M.:* Cell surface changes during muscle differentiation *in vitro;* a study with the probe 2,4,6-trinitrobenzene sulfonate. Devl Biol. (submitted 1977).
15 *Schiaffino, S.; Cantini, M., and Sartore, S.:* T-system formation in cultured rat skeletal muscle. Tissue Cell (in press).
16 *Schubert, D.; Harris, A.J.; Heinemann, S.; Kidokoro, Y.; Patrik, J., and Steinbach, J.H.:* Differentiation and interaction of clonal cell lines of nerve and muscle; in *Stato* Tissue culture of the nervous system, pp. 55–85 (Plenum Press, New York 1975).
17 *Sisto Daneo, L.:* Inattesa precocità della differenziazione delle fibre muscolari lente e rapide. 34th Congr. Soc. Ital. Anat., Trieste 1977.
18 *Sisto Daneo, L. and Filogamo, G.:* Differentiation of synaptic area in slow and fast muscle fibres. J. submicrosc. Cytol. *7:* 121–131 (1975).

Prof. *G. Filogamo,* Istituto di Anatomia Umana Normale dell'Università di Torino, Corso Massimo d'Azeglio 52, *I–10100 Torino* (Italy)

Formation of the Neuronal Circuitry of Rat Cerebellum: Plasma Membrane Modifications

G. Gombos, M.S. Ghandour, G. Vincendon, A. Reeber and J.-P. Zanetta

Centre de Neurochimie du CNRS et Institut de Chimie Biologique de la Faculté de Médecine, Université Louis Pasteur, Strasbourg

Introduction

The hypotheses concerning the roles that the carbohydrate moiety (glycan) of glycoconjugates of cell surface, in particular glycoproteins, can play in processes of cell growth, differentiation, intercellular recognition and establishment of connections, hence in tissue morphogenesis (for reviews, see 14, 38), rest mainly on the fact that glycans of plasma membranes are the most peripheral compounds on cell surface (35) and that they have complex structures; this complexity is at the basis of their alleged 'coding' properties (38). Different patterns of distribution of different glycans on the cell surface and their changes at different stages of cellular development could play a role as determinants of cell surface specificity (of the whole plasma membrane and of its different areas on the same cells) and of developmental modifications of such specificity.

The scarce data obtained with cells derived from nervous tissue are consistent with (although they are not a proof of) these hypotheses. Carbohydrate material accumulates at the surface of the growth cone of some neuroblasts (21, 33). The composition of cell surface glycoproteins of neuroblastoma cells is modified by neurite sprouting (10, 18, 42) and that of the bulk of brain glycoproteins changes throughout development (16, 22, 25, 26). The use of labelled lectins (sugar-binding proteins) of different specificity, in particular of Con A[1], has shown not only that the distribution of glycans which bind to one lectin is not uniform on the whole surface of one type of neuroblast and different in different neuroblasts, but also that neuroblasts of different types

[1] Concanavalin A (Con A) is a lectin which binds *D*-mannose and/or glucose provided the hydroxyls linked to carbons 3, 4 and 6 of these sugars are free, that is, only if these sugars are in terminal position in a polysaccharide, or internal and linked through carbons 1 and 2 (19).

have different patterns of lectin binding sites on their surface (33, 34, 43). However, direct experimental evidence that glycans of cell surface glycoproteins play, as suggested (7, 11), a role in nervous tissue morphogenesis, and in the establishment of specific cell connections, hence in the formation of neuronal circuits, has not yet been obtained. We have attempted to detect, in the CNS *in situ,* developmental changes of these glycans in a given neuronal type and to characterize these changes at molecular level. We hoped to identify those glycan-carrying molecules which could interact, at a precise phase of differentiation of one neuroblast, with specific molecules on the surface of another neuroblast with which the first should make synapses (partner cell). These molecules on the partner cell surface could be lectin-like molecules such as those described in slime molds, electric organ, muscle, cultured myoblasts, neuroblastoma cells, probably chick brain (8, 30, 41) and more recently, in developing rat brain (39).

Our study was carried out on rat cerebellum, chosen because of its morphology and its development (see below), and above all, because the overwhelming predominance of one cell type made feasible the correlation between biochemical and morphological data. In fact, in the adult cerebellar cortex (folium), neuronal heterogeneity is limited to six, easily recognizable cell types, five of which are interneurons and one, the granule cell, accounts for 90% (13) of the cells. Neurons are arranged into three well-defined layers (molecular, ganglionar or Purkinje cell and granular layers (31, 36) connected into identical circuits throughout the folium (17, 24), through processes which can be identified simply on morphological criteria. The small perikarya and short dendrites of granule cells are all packed together in the granular layer, while their long axons, after crossing the folium, become the parallel fibers, which, all packed together in homogeneous bundles, account for most of the molecular layer volume (31, 32, 36). This has allowed us to estimate that the overall surface of parallel fibers is between 50 and 90 times greater than that of granule cell perikarya (32, 47), hence the axolemma of parallel fibers is the predominant fraction of plasma membranes of the adult cerebellar folium.

Cell Multiplication and Membrane Formation during Postnatal Development of Rat Cerebellum

The morphology and chronology of differentiation of each neuron type in rat cerebellar folium has been described in detail (1–3), and at present it is not difficult to identify neuroblasts that differentiate into each neuron type and to follow their migration toward their final position and the outgrowth of their processes.

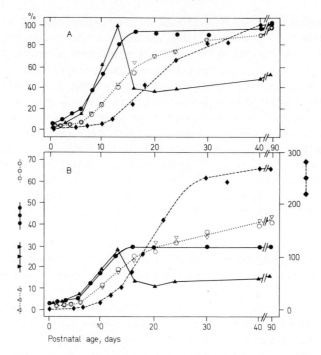

Fig. 1. Postnatal development of rat cerebellum. Age-dependent modifications, on a per cerebellum basis of dry weight (○); ouabain-inhibited, Na$^+$, K$^+$-dependent ATPase (♦); DNA (●), and protein in total cerebellar homogenate (▽) as compared to the developmental curve of membrane-bound, Con A-binding glycoproteins in 'crude membrane fraction' (▲). Ordinates: *A* Amount at each age expressed as percent of the maximum amount attained during development. *B* Amount at each age/amount at birth.

The cerebellum of a newborn rat contains only about 3% of the DNA of an adult cerebellum (5, 13) (fig. 1) and its neuronal population consists of neurons of the deep nuclei and of the neuroblasts which are differentiating into Purkinje cells and Golgi interneurons (1–3). The other interneurons of the folium (granule, stellate, basked and, possibly, Lugaro cells) all derive from the postnatal proliferation of stem cells of the external germinative layer (EGL) (1–3). Such proliferation is relatively slow for the first postnatal week, is extremely rapid until the 16th day and is terminated by the 21st day (3, 5, 12) (fig. 1). As we have discussed elsewhere (47), already at the end of the first postnatal week, the bulk of cells formed are granule cells. During the 2-day migration of each newly formed granule cell from the EGL to the granular layer (1–3), the granule cell axon grows at one end by following the migrating perikaryon and at the other end by forming the parallel fiber. Each newly formed parallel fiber piles up above older parallel fibers. The packing of parallel fibers is spread over the same time span as granule cell proliferation and to few additional days until by the 30th postnatal day all parallel fibers are in place (1–3) and have made synapses with the dendrites of Purkinje, Golgi, basket and stellate neurons (17, 31). At this date, all other cerebellar neurons and their circuitry have the same morphology as in the adult (1–3) but biochemical differentiation appears to continue up to the 40th day. In fact, adult values, on a per cerebellum basis, of dry weight, protein (42–43 times as high as in the newborn) and of the plasma membrane marker Na^+-K^+ ATPase, are not attained before this date (fig. 1). The high increase of the enzyme activity (265-fold the value in newborn) (fig. 1) is another indication of the exceptional development of cell processes, hence of plasma membranes in cerebellum if we assume that the enzyme activity is roughly proportional to plasma membrane surface.

Con A-Binding Sites during Cerebellar Development

Cerebellar slices of different ages were incubated with lectins of different specificity labelled either with fluorescein isothiocyanate (FITC) or by the horseradish peroxidase method (HRP) (see *Zanetta et al.*, 47, for detailed description). The most impressive results were obtained when we used Con A (47) (fig. 2). Although many structures specifically bind the lectin (44, 47) and white matter also shows a nonspecific binding (47), nowhere are specific Con A binding and its developmental changes so striking as in the molecular layer (47): massive binding of Con A to the molecular layer is a transient phenomenon during development (fig. 2) and is due to the presence of abundant Con A-binding sites on the axolemma of growing parallel fibers (fig. 3). The phenomenon is particularly evident also because the Con A-binding sites on the tightly packed bundles of parallel fibers are topographically more concentrated than those on other, dispersed, structures.

The overwhelming preeminence of parallel fiber axolemma on other cerebellar membranes allowed the identification, in 'crude membrane fractions'[2], of the transient Con A-binding sites: among all Con A-binding molecules, only membrane-bound Con A-binding glycoproteins accumulate and then decrease with the same time course as the transient Con A-binding sites on parallel fibers

---

[2] Crude membrane fractions were the $100,000\,g$ 120 min pellet of cerebellar homogenates in hypotonic buffers (100 vol of 20 mM Tris-HCl, pH 7.2) (47). After lipid extraction, the proteins of this fraction were solubilized in SDS and the sulfhydryl groups alkylated as described (20). Con A-binding glycoproteins were separated from the bulk of proteins of this fraction by affinity chromatography on Con A-Sepharose as previously described (45, 46). Polyacrylamide gel electrophoresis, in the presence of SDS (PAGE–SDS), was carried out in acrylamide concentration gradients (4–30%) gel slabs.

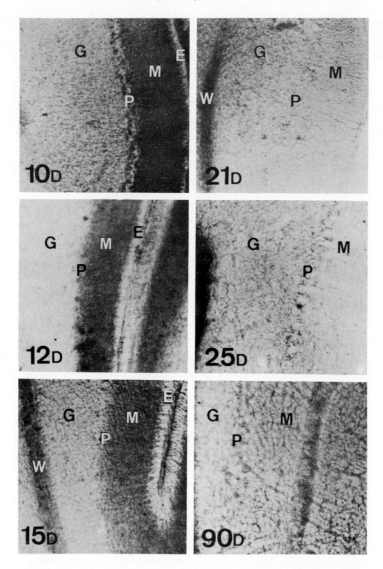

Fig. 2. Parasagittal sections of unstained rat cerebellum. Con A-binding sites were revealed, as described (47), by the Con A-peroxidase method. 10D, 12D, 15D, 21D, 25D and 90D = animal postnatal age in days; G = granular layer; P = Purkinje cell layer; M = molecular layer; E = external granular (germinative) layer; W = white matter. The strong labelling of white matter at all ages is not removed by α-methyl-mannoside, specific haptenic inhibitor of Con A. On the contrary most or all of the staining of the molecular layer and of other structures is removed. × 100 (at the microscope; reduced for publication). From Zanetta et al. (47).

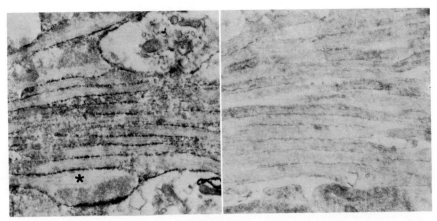

Fig. 3. Electron micrographs of immature parallel fibers in the molecular layer of the cerebellar cortex in a 12-day-old rat. Section along the length of parallel fibers. Unstained sections, only postfixed with osmic acid, were processed with the HRP-Con A procedure. *A* The product of HRP-Con A reaction is concentrated around the plasma membrane of parallel fibers and of a structure containing synaptic vesicles (*). *B* No product of HRP-Con A reaction is detected when sections were incubated with Con A in the presence of α-methyl-mannoside. × 17,500 (at the microscope; reduced for publication). From *Zanetta et al.* (47).

(47). PAGE-SDS show that the steady and slow increase (see dotted line on fig. 4) of Con A-binding glycoproteins, until they account for about 3% of cerebellar proteins (47), is due to the accumulation of many glycoproteins which appear at different stages of cerebellar development, but the peak, which at its maximum accounts for 13–15% of cerebellar membrane proteins (fig. 1, 4) (47), is mostly due to the accumulation of only four glycoprotein subunits (*Zanetta et al.*, unpublished).

These subunits decrease and become undetectable from the 20th postnatal day on, simultaneously to the net decrease, on per cerebellum basis, of Con A-binding glycoproteins (fig. 4). This strongly suggests that the transient Con A-binding sites on parallel fibers correspond to these glycoprotein subunits, and thus the sites on parallel fibers disappear because the glycoproteins either disappear or lose affinity for Con A (see below) and not because their glycans become 'masked' by contiguous molecules nor because of a redistribution of sites on the axolemmal surface. The chronological relationship between disappearance of Con A-binding sites and formation of abundant synapses between parallel fibers and dendrites of Purkinje, basket, stellate and Golgi neurons and the abundance of Con A-binding sites in synapses (9, 15, 27) could have suggested that Con A-binding sites diffused on the axolemmal surface have moved to form discrete patches in the areas where synapses were to be formed.

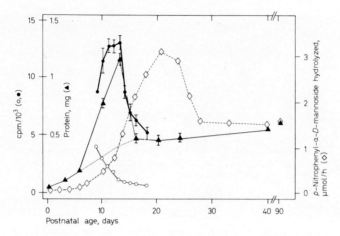

Fig. 4. Postnatal development of rat cerebellum. Age-dependent modifications, on a per cerebellum basis ± SD, of Con A-binding glycoproteins in the 'crude membrane fraction' (▲), and of α-mannosidase in the total cerebellar homogenate (◇) as compared to the radioactivity curve of TCA soluble (○) and TCA insoluble (●) fractions of whole cerebellum homogenates after intraperitoneal injection of U-^{14}C-glucosamine at the 8th postnatal day.

However, on the basis of the data presented, it must be concluded that either the glycoproteins corresponding to the transient Con A-binding sites on parallel fiber do not become structural constituents of synapses, and thus they are not involved in the maintenance of the junction, or if they do, they have lost affinity for Con A.

Possible Mechanism of Disappearance of Con A-Binding Sites on Growing Parallel Fibers

The disappearance of some Con A-binding glycoproteins is not necessarily due to a specific breakdown of these molecules but could be simply due to loss of affinity for Con A because of glycan modifications. This hypothesis implies that some glycoproteins synthesized during the whole life of the granule have different glycans at different periods of development. One possibility *(glycan completion hypothesis)* is that Con A-binding sites are 'core-like' (28) incomplete glycans, with mannose as terminal sugar. These glycans are either completed, *in situ,* by the addition of other sugars (mediated by glycosyltransferases which appear, at later ages on parallel fiber axolemma?) or substituted, because of membrane turnover, by glycoproteins with the same protein moiety but with complete glycans. In the last case, we must assume that young granule cells are incapable of synthesizing complete glycans. The 'glycan completion hypothesis'

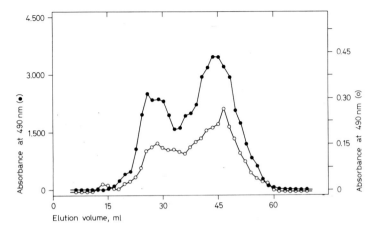

Fig. 5. Comparison, by gel filtration on Biogel P-30, of molecular size of glycopeptides derived from 'crude membrane fractions' of adult (●) and of 11-day-old (○) rat cerebella. Ordinates: absorbance at 490 nm of the chromogen developed with the phenol-sulfuric acid method for hexose detection. Glycopeptides were obtained by pronase digestion and purified as described elsewhere (37). The Biogel P-30 column (26 × 2 cm) was equilibrated and eluted with 50 mM ammonium formate buffer pH 7.0 containing 0.02% sodium azide.

is unlikely since the distribution of cerebellar glycoprotein glycans, according to molecular weight, is the same in 11-day-old as in adult cerebella (fig. 5) while it should have been expected that 'core-like' glycans have smaller molecular weights than complete glycans, and thus, with age, lesser amounts of small molecular weight glycans should be present. It is also unlikely since labelled lectins specific for peripheral sugars in complex glycans (*Ricinus* lectin for terminal galactose, UEL for *L*-fucose extracted from *Ulex europeus* seeds) and WGA (specific for N-acetylglucosamine) never showed later during development any phenomenon similar to those obtained with Con A as should have been the case if the loss of affinity to Con A was due to the addition of more peripheral sugars (Zanetta et al., unpublished). Furthermore is it also unlikely since granule cells at early ages and even EGL cells are able to synthesize complex glycans as shown by the binding of UEL to their plasma membranes (Zanetta et al., unpublished).

Thus the alternate hypothesis, that of 'glycan breakdown', is more probable. In favor of this hypothesis is the curve of the radioactivity derived from U-^{14}C-glucosamine incorporated into cerebellar proteins. In adult cerebella, the radioactivity decay curve follows a gentle slope. On the contrary, in animals injected at the 8th postnatal day, the same radioactivity shows a sudden drop between the 14th and the 16th postnatal day (fig. 4). This abrupt decrease suggests a rapid degradation of a large class of cerebellar glycoprotein glycans.

Associated with the decrease of Con A-binding glycans, and alone among cerebellar glycosidases, is the transient increase of α-mannosidase (fig. 4). These data however do not show if the whole glycoprotein molecule is broken down or only its glycans. However, the systematic removal of specific Con A-binding glycans after the 20th postnatal day and throughout life is unlikely and thus, all the circumstantial evidence suggests that some specific glycoproteins are synthesized only during a phase of granule cell maturation and that their disappearance at a more advanced age is due to breakdown of the whole molecule (see below).

Possible Biological Role of Transient Con A-Binding Sites

Although the transient Con A-binding sites in parallel fibers do not appear to be constituents of synapses, they could still be involved in processes preliminary to the formation of the junction such as interneuronal recognition and 'early' connection. This 'early' connection could consist of interaction of some glycans on the axolemma of parallel fibers with (lectin-like?) molecules on the surface of the postsynaptic partner cell of parallel fibers. In order to verify this hypothesis, Con A-binding glycoproteins from 'crude membrane fractions' of 15-day-old cerebella were solubilized and fractionated by sequential treatment with detergents, then they were either made fluorescent with dansyl-chloride or coupled to HRP, finally detergents were removed. A glycoprotein fraction, now soluble in absence of detergent, appeared to bind to perikarya and dendrites of all cells with which parallel fibers make synapses. Such binding was present only at ages at which Con A-binding glycoproteins were present on parallel fibers.

Thus we postulate that some specific glycoproteins are very abundant on growing parallel fibers at the same time that their 'complementary' molecules are abundantly present on the whole plasma membrane of the postsynaptic partner cells. The molecular redundance of the two types of complementary molecules on the two compatible surfaces should increase the probability for the meeting of these molecules and, thus, for recognition of, and interaction between, partner cells. Since the Con A-binding glycoproteins are denatured by the detergent treatment, it appears that their configuration is not important and, thus, that probably only their glycans are involved in the interaction with 'the complementary' molecules on the partner cells. The establishment of the 'early' connection should be followed with the 'construction of the synapses' which probably, from this time on, is independent of cellular specificity. The disappearance of the transient Con A-binding sites on parallel fiber axolemma simultaneous to synapse construction could be genetically programed or even induced by synapse formation.

Desynchronization of Purkinje cell maturation and granule cell formation by thyroid hormones (23, 29) or X-ray (for other references, see 4, 6) should confirm or reject these hypotheses. Similarly the absence of either group of complementary molecules could be at the basis of the absence of synapses between parallel fibers and Purkinje cell dendrites and of the transsynaptic degeneration 'en cascade' of granule cells in staggerer mutant mice (40).

Summary

During morphogenesis of rat cerebellum, the axolemma of the growing axons (parallel fibers) of granule cells are coated by Con A-binding sites which disappear as parallel fibers form synapses with the dendrites of their 'partner' cells. These sites consist of a few membrane-bound Con A-binding glycoprotein subunits. Circumstantial evidence suggests that the transient character of these molecules is due to their disappearance rather than to modification of their glycans with consequent loss of affinity for Con A. Sites which bind these glycoproteins are also transiently present on the whole plasma membrane of the postsynaptic partner cells of parallel fibers.

We suggest that both, Con A-binding glycoproteins in parallel fiber axolemma and their binding sites on the postsynaptic partner cell plasma membrane are involved in the phase of cell recognition which precedes the establishment of appropriate synapses. Two factors are very important in these phenomena, one is the redundance of the two groups of molecules, the other is the time course of their appearance and disappearance, which is not exactly the same for the two groups of molecules.

Acknowledgements

We thank Ms. *O. Levy, A. Meyer, R. Langs* and *Y. Schladenhaufen* for their skilful technical assistance. This work was partially supported by a grant from the Centre National de la Recherche Scientifique (ATP 'Génétique et Développement d'un Mammifère' décision No. 2374). *G. Gombos* and *J.-P. Zanetta* are respectively Maître and Chargé de Recherche au CNRS.

References

1 *Altman, J.:* Postnatal development of the cerebellar cortex in the rat. I. The external germinal layer and the transitional molecular layer. J. comp. Neurol. *145:* 353–398 (1972).
2 *Altman, J.:* Postnatal development of the cerebellar cortex in the rat. II. Phases in the maturation of Purkinje cells and of the molecular layer. J. comp. Neurol. *145:* 399–464 (1972).
3 *Altman, J.:* Postnatal development of the cerebellar cortex in the rat. III. Maturation of the components of the granular layer. J. comp. Neurol. *145:* 465–514 (1972).
4 *Altman, J.:* Experimental reorganization of the cerebellar cortex. VII. Effect of late X-irradiation schedules that interfere with cell acquisition after stellate cells are formed. J. comp. Neurol. *165:* 65–76 (1976).

5   Balázs, R.; Kovacs, S.; Cocks, W.A.; Johnston, A.L., and Eayrs, J.T.: Effect of thyroid hormone on the biochemical maturation of rat brain: postnatal cell formation. Brain Res. 25: 555–570 (1971).
6   Balázs, R.; Lewis, P.D., and Patel, A.J.: Effects of metabolic factors on brain development; in Brazier Growth and development of the brain, pp. 83–115 (Raven Press, New York 1975).
7   Barondes, S.H.: Brain glycomacromolecules and interneuronal recognition; in Schmitt The neurosciences: second study program, pp. 747–760 (Rockefeller University Press, New York 1970).
8   Barondes, S.H. and Rosen, S.D.: Cell surface carbohydrate-binding proteins. Role in cell recognition; in Barondes Neuronal recognition, pp. 331–356 (Plenum Press, New York 1976).
9   Bittiger, H. and Schnebli, H.P.: Binding of concanavalin A and ricin to synaptic junctions of rat brain. Nature, Lond. 249: 370–371 (1974).
10 Brown, J.C.: Surface glycoprotein characteristic of the differentiated state of neuroblastoma C-1300 cells. Expl. Cell Res. 69: 440–442 (1971).
11 Brunngraber, E.G.: Possible role of glycoproteins in neural function. Perspect. Biol. Med. 12: 467–470 (1969).
12 Chanda, R.; Woodward, D.J., and Griffin, S.: Cerebellar development in the rat after early postnatal damage by methylazoxymethanol: DNA, RNA and protein during recovery. J. Neurochem. 21: 547–555 (1973).
13 Clos, J.; Fabre, C.; Seime-Matrat, M., and Legrand, P.: Effect of undernutrition on cell formation in the rat brain and specially on cellular composition of the cerebellum. Brain Res. 123: 13–26 (1977).
14 Cook, G.M.W. and Stoddart, R.W.: Surface carbohydrates of the eukaryotic cell, pp. 257–270 (Academic Press, London 1973).
15 Cotman, C.W. and Taylor, D.: Localization and characterization of concanavalin A receptors in the synaptic cleft. J. Cell Biol. 62: 236–242 (1974).
16 Di Benedetta, C. and Cioffi, L.A.: Glycoproteins during the development of the rat brain; in Zambotti, Tettamanti and Arrigoni Glycolipids, glycoproteins and mucopolysaccharides of the nervous system, pp. 115–124 (Plenum Press, New York 1972).
17 Eccles, J.C.; Ito, M., and Szentágothai, J.: The cerebellum as a neuronal machine (Springer, Berlin 1967).
18 Glick, M.C.; Kimhi, Y., and Littauer, U.Z.: Glycopeptides from surface membranes of neuroblastoma cells. Proc. natn. Acad. Sci. USA 70: 1682–1687 (1973).
19 Goldstein, I.J.: Carbohydrate binding specificity of concanavalin A; in Bittiger and Schnebli Concanavalin A as a tool, pp. 55–65 (Wiley, London 1976).
20 Gombos, G.: Solubilization of brain membranes for affinity chromatography of glycoproteins; in Bittiger and Schnebli Concanavalin A as a tool, pp. 379–387 (Wiley, London 1976).
21 James, D.W. and Tresman, R.L.: The surface coats of chick dorsal root ganglion cells in vitro. J. Neurocytol. 1: 383–395 (1972).
22 Krusius, T.; Finne, J.; Kärkkäinen, J., and Järnefelt, J.: Neutral and acidic glycopeptides in adult and developing rat brain. Biochim. biophys. Acta 365: 80–92 (1974).
23 Lewis, P.D.; Patel, A.J.; Johnson, A.L., and Balázs, R.: Effect of thyroid deficiency on cell acquisition in the postnatal rat brain: a quantitative histological study. Brain Res. 104: 49–62 (1976).
24 Llinás, R. and Hillman, D.E.: Physiological and morphological organization of the cerebellar circuits in various vertebrates; in Llinás Neurobiology of cerebellar evolution and development, pp. 43–73 (Am. Med. Ass./Education & Research Found., 1969).

25 *Margolis, R.K. and Gomez, Z.:* Structural changes in brain glycoproteins during development. Brain Res. *74:* 370–372 (1974).
26 *Margolis, R.K.; Preti, C.; Lai, D., and Margolis, R.U.:* Developmental changes in brain glycoproteins. Brain Res. *112:* 363–369 (1976).
27 *Matus, A.; Petris, S. de, and Raff, M.C.:* Mobility of concanavalin A receptors in myelin and synaptic membranes. Nature new Biol. *244:* 278–279 (1973).
28 *Montreuil, J.:* Recent data on the structure of the carbohydrate moiety of glycoproteins. Metabolic and biological implications. Pure appl. Chem. *42:* 431–477 (1975).
29 *Nicholson, J.L. and Altman, J.:* The effects of early hypo- and hyperthyroidism on the development of rat cerebellar cortex. II. Synaptogenesis in the molecular layer. Brain Res. *44:* 25–36 (1972).
30 *Nowak, T.P.; Haywood, P.L., and Barondes, S.H.:* Developmentally regulated lectin in embryonic chick muscle and a myogenic cell line. Biochem. biophys. Res. Commun. *68:* 650–657 (1976).
31 *Palay, S.L. and Chan-Palay, V.:* Cerebellar cortex cytology and organization (Springer, Berlin 1974).
32 *Palkovitz, M.; Magyar, P., and Szentágothai, J.:* Quantitative histological analysis of the cerebellar cortex in the cat. III. Structural organization of the molecular layer. Brain Res. *34:* 1–18 (1971).
33 *Pfenninger, K.H. and Maylie-Pfenninger, M.F.:* Distribution and fate of lectin binding sites on the surface of growing neuronal processes. J. Cell Biol. *67:* 332a (1975).
34 *Pfenninger, K.H. and Rees, R.P.:* From the growth cone to the synapse. Properties of membranes involved in synapse formation; in *Barondes* Neuronal recognition, pp. 131–173 (Plenum Press, New York 1976).
35 *Rambourg, A. and Leblond, C.P.:* Electron microscope observations on the carbohydrate-rich cell coat present at the surface of cells in the rat. J. Cell Biol. *32:* 27–53 (1967).
36 *Ramón Y Cajal, S.:* Histologie du système nerveux de l'homme et des vertébrés, vol. 2 (Maloine, Paris 1911).
37 *Reeber, A.; Zanetta, J.-P.; Morgan, I.G., and Gombos, G.:* Purification and analysis of glycopeptides derived from nervous tissue membranes; in Méthodologie de la structure et du métabolisme des glycoconjugués. Coll. Int. No. 221 CNRS, Villeneuve d'Ascq 1974, vol. II, pp. 815–828.
38 *Sharon, N.:* Complex carbohydrates. Their chemistry, biosynthesis and function, pp. 26–29, 177–190 (Addison-Wesley, Reading, Mass. 1975).
39 *Simpson, D.L.; Thorne, D.R., and Loh, H.H.:* Developmentally regulated lectin in neonatal rat brain. Nature, Lond. *266:* 367–369 (1977).
40 *Sotelo, C. and Changeux, J.P.:* Transsynaptic degeneration 'en cascade' in the cerebellar cortex of staggerer mutant mice. Brain Res. *67:* 519–526 (1974).
41 *Teichberg, V.I.; Silman, I.; Beitsch, D.D., and Resheff, G.:* A β-D-galactoside binding protein from electric organ tissue of *Electrophorus electricus.* Proc. natn. Acad. Sci. USA *72:* 1383–1387 (1975).
42 *Truding, R.; Shelanski, M.L.; Daniels, M.P., and Morell, P.:* Comparison of surface membranes isolated from cultured murine neuroblastoma cells in the differentiated or undifferentiated state. J. biol. Chem. *249:* 3973–3982 (1974).
43 *Vaughn, J.E.; Henrikson, C.K., and Wood, J.G.:* Surface specializations of neurites in embryonic mouse spinal cord. Brain Res. *110:* 431–445 (1976).
44 *Wood, J.G.; McLaughlin, B.J., and Barber, R.P.:* The visualization of concanavalin A-binding sites in Purkinje cell somata and dendrites of rat cerebellum. J. Cell Biol. *63:* 541–549 (1974).

45 Zanetta, J.-P.; Morgan, I.G., and Gombos, G.: Synaptosomal plasma membrane glycoproteins: fractionation by affinity chromatography on concanavalin A. Brain Res. *83:* 337–348 (1975).
46 Zanetta, J.-P. and Gombos, G.: Affinity chromatography of brain membrane glycoproteins on concanavalin A-Sepharose in the presence of SDS; in *Bittiger and Schnebli* (Concanavalin A as a tool, pp. 389–398 (Wiley, London 1976).
47 Zanetta, J.-P.; Roussel, G.; Ghandour, M.S.; Vincendon, G., and Gombos, G.: Postnatal development of rat cerebellum: massive and transient accumulation of concanavalin A-binding glycoproteins in parallel fiber axolemma. Brain Res. *140* (in press, 1978).

Dr. *G. Gombos,* Centre de Neurochimie du CNRS, Faculté de Médecine, 11, rue Humann, *F–67085 Strasbourg Cedex* (France)

Neuromuscular Alterations as a Consequence of Cholinergic Blockade during Development[1]

Giacomo Giacobini

Department of Human Anatomy, University of Turin, Turin

Embryonic muscles provide an excellent material for studies concerned with cell interactions during the development of cholinergic synapses. In comparison with other cholinergic districts (such as sympathetic ganglia or discrete brain and spinal cord areas), they show the advantage of a larger size and of a relative homogeneity of their synaptic population. A further advantage is represented by the nonnervous nature of the postsynaptic element (lacking transmitter systems other than that of the neuromuscular synapse itself). Thus, on the basis of the pioneering morphological investigations of *Cajal* (3), the normal development of neuromuscular correlations has been exhaustively studied in the last 15 years from the ultrastructural, histochemical and biochemical points of view; some physiological data are also available. As a result of all these researches, we have at present a relatively detailed picture of the development of these synapses in many vertebrates, especially in the chick, which provides the most suitable material for developmental studies.

The normal development of neuromuscular correlations follows two main phases (6, 7). During the first phase (3rd to 10th/11th day in the chick embryo; for pertinent literature, see *Filogamo,* 5), root motor fibers, growing out of the spinal cord, reach myotomes and start branching among myoblasts; they release acetylcholine (ACh), and both acetylcholinesterase (AChE) and cholinergic receptor (ACh-R) are present at this stage in myoblasts. During this period, the fusion of myoblasts into myotubes takes place. At the beginning of the second phase (10th/11th day), motor end-plates appear (6, 13, 14), which will reach functional maturity before hatching (1); this is accompanied by the synthesis of increasing amounts of choline acetyltransferase (ChAc) by the motoneuron (2, 8). Myotubes transform into true muscle fibers, and AChE and ACh-R are almost exclusively localized at end-plates.

[1] Supported by grants from the Italian National Research Council No. 75.00580.04, 76.01245.04, and 76.01398.04.

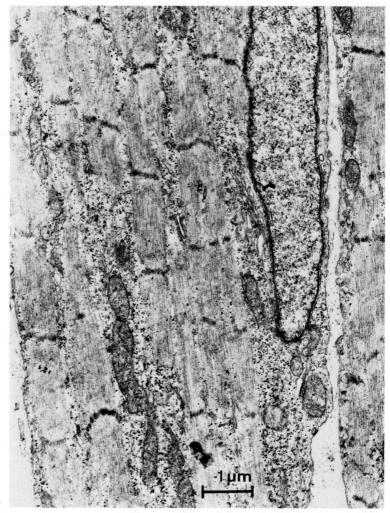

Fig. 1. Electron micrographs of skeletal (leg) muscles from 16-day-old chick embryos. *a* Normal embryo; note regularly organized myofilaments. *b* Botulinum toxin treated embryo. Two myotubes at different degrees of maturation; note centrally located nuclei.

The presence of the components of the ACh system, ChAc, AChE and ACh-R during the first step (6, 8, 9) suggests that cholinergic transmission is already functional in the absence of well-defined synaptic structures and confirms that movements are neurogenic from their very beginning at the 3rd day, as first suggested in the 30s by *Visintini and Levi-Montalcini* (15). During the second phase, end-plate formation is accompanied by a preferential synthesis of

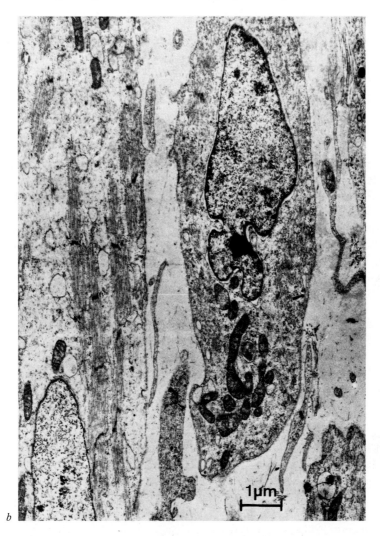

b

the proteins of the postsynaptic membrane, AChE and ACh-R, in respect of total muscular proteins.

Owing to the availability of specific cholinergic blocking agents (acting at pre- and postsynaptic levels), the reciprocal influences of nervous and muscular elements have been experimentally investigated during *in vivo* development. Our experimental approach to the problem of neuromuscular synapse formation is mainly based on the action of two toxins: *Naja nigricollis* α-toxin, which binds to nicotinic ACh-R, and type A botulinum toxin, which blocks release of ACh by nerve terminals (9–11). Toxins were injected at given times in the yolk sac, through a small window in the shell. Treatment started very early, generally at the 3rd day of incubation.

Both toxins produce a marked regression of body weight, essentially due to muscle atrophy. As already observed by *Drachman* (4), who first employed botulinum toxin to study the influence of motor nerves on the developing muscle, the degree of atrophy increases with age, mainly after the 12th day (it must be emphasized nevertheless that some effects may be observed before this stage). In our experimental work, the dry weight of treated muscles was about 50% of control at the 16th day. At this stage, control muscles were mainly represented by true muscle fibers with regularly arranged myofilaments (fig. 1a); treated muscles contained many myoblasts and myotubes (fig. 1b); the injection schedule is important as far as effects on muscle are concerned. When toxin treatment started at the 3rd day, in particular a delay in maturation was observed, and many myoblasts were present. When injection started at the 7th day, degenerative phenomena (for example, disorganization of myofilaments) were frequent, especially in more mature elements (myotubes or even true muscle fibers). It seems likely that all these effects of delayed differentiation and degeneration are determined by the lack of activation of the postsynaptic membrane and, therefore, of muscle contraction. The action on muscle of both α-toxin and botulinum toxin results thus in a pharmacological cholinergic denervation; indeed, most of the effects observed resemble those seen after surgical denervation in the chick embryo (for discussion of the neurogenic control of muscle differentiation, see *Drachman*, 4; *Giacobini et al.*, 9, and *Giacobini-Robecchi et al.*, 10).

When the motor innervation of treated muscles is considered, some differences may be observed between the effects of the two toxins. The effect of botulinum toxin was very low (10). Many nerve fibers were detected within muscles, contacting even immature elements. Thus, a large number of very simple synapses was observed, mainly on myotubes and myoblasts. ChAc, taken as a presynaptic marker, was practically unchanged in muscles and in nerves; the postsynaptic proteins AChE and ACh-R, on the contrary, were markedly reduced.

In embryos treated with α-toxin, marked presynaptic effects may be observed, in spite of the postsynaptic site of action of the toxin itself. In a previous series of results (9) we noticed a high reduction of motor end-plates and a fall not only in AChE, but also in ChAc activity in muscles. Since in muscles ChAc is unique to motor fibers and terminals (12), a fall in its activity would indicate damage to the motor innervation of the muscle itself. These preliminary results raised a question: how can the motoneuron be damaged or its develop-

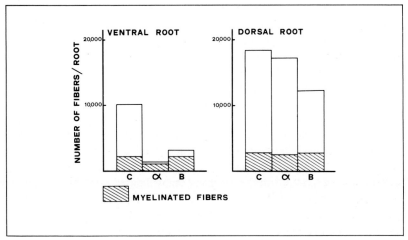

Fig. 2. Effects of chronic α-toxin and botulinum toxin treatment on the number of fibers in ventral and dorsal roots of the 4th lumbar nerve (16-day-old chick embryos). C = Control; α = α-toxin; B = botulinus toxin. Based on data from *Giacobini-Robecchi et al.* (11).

ment be affected in consequence of a postsynaptic blockade? In other words, how can the blocked muscle inform the motoneuron of its state? Since we had reasons to exclude a transport of information through sensory fibers (see below), we were led to postulate the existence of a retrograde signal at the level of the neuromuscular synapse itself. Information would then pass directly from the postsynaptic to the presynaptic cell in the opposite direction to the transmission of impulses. In order to clarify this point, we have undertaken a series of further investigations concerning the effects of α-toxin on motor innervation.

First of all, we confirm that the number of end-plates is greatly reduced in α-toxin treated embryos. In most atrophic specimens, only a few exceptional end-plates could be detected by the Koelle method for AChE; electron microscope observations revealed a very simple structure. Moreover, nerve fibers detected both by light and electron microscopy are extremely rare within muscles. The high reduction of end-plates is consistent with the fall in the activities of the enzymes of the ACh system: the presynaptic enzyme ChAc is significantly reduced both in muscles (as already stated) and in peripheral (sciatic) nerve. The reduction of AChE activity is in agreement with histochemical observations using the Koelle method.

In order to confirm the selective damage of motor elements after α-toxin treatment, we have undertaken fiber number counts on electron micrographs of ventral and dorsal roots of a spinal (4th lumbar) nerve in control and experimental embryos (fig. 2) (11). Dorsal roots of control 16-day-old embryos contain a mean of about 18,000 fibers per root (since at this stage myelination is not yet completed, only about 15% of the fibers are myelinated). A nonsignificant loss of nonmyelinated fibers has been noticed in α-toxin treated embryos, and a decrease of less than 30% in botulinum treated ones. Ventral roots of control embryos contain about 10,000 fibers per root (about 2,000, or 20%, with myelin sheath). In botulinum toxin treated embryos, a decrease to 30% of the control, almost exclusively in the number of nonmyelinated fibers, has been measured. The α-toxin treated embryos show maximal reduction: myelinated fibers are about 50% of the control and nonmyelinated ones are practically absent. Thus, total fiber population is reduced to 13% of control. This remarkable effect of α-toxin on ventral roots is immediately evident at electron microscope observation: in the control, nonmyelinated fibers are very abundant, collected in bundles among myelinated ones. Following chronic α-toxin treatment, nonmyelinated fibers almost disappear, and many of the myelinated fibers are enlarged or show more or less advanced degeneration processes, often with good preservation of the myelin sheath.

These results confirm that presynaptic effects are observed mainly after α-toxin treatment, and that only the motor elements of the reflex arc are significantly affected. The selectivity of action of the α-toxin on motor elements (whatever the mechanism involved) was controlled also in the spinal cord: while dorsal horn neurons seemed to be unaffected, motoneurons were reduced in number and showed signs of degeneration with vacuoles and disorganization of Nissl bodies.

In conclusion, the presynaptic effects of a toxin acting at the postsynaptic level, like α-toxin, suggest that a transfer of informations occurs during normal development from muscle cells to motoneurons. The blockade of the ACh-R, and thus of neuromuscular transmission and muscle contraction, abolishes the information which the developing motoneuron requires to correctly mature and to maintain its structural and functional integrity. The question is raised concerning the nature of this information and the pathway followed by it to reach the motoneuron. The sensory pathway from muscle to spinal cord is unlikely to be involved (for discussion, see *Giacobini et al.,* 9) and sensory elements seem not to be affected in our experimental embryos (11). A specialized signal seems more likely to be transferred directly through the synapse itself, in a direction opposite to that of impulse transmission. The nature of this hypothetical retrograde signal is not known; we proposed that it could consist of a low concentration of calcium ions in the nerve terminal (10). It is possible, however, that other factors, such as, for example, a small protein similar to the nerve growth factor could be involved in this process.

Whatever the chemical nature of this hypothetical factor, it would then control the stabilization of the developing synapse and the survival of the presynaptic element. This stabilized synapse would thus be allowed in its turn to intervene in maintaining the integrity of the postsynaptic cell. The maturation of neuromuscular correlations seems thus to be critically regulated by mutual collaborations between pre- and postsynaptic elements, that is between center and periphery, during development.

Summary

The action of some cholinergic blocking agents (snake α-toxin, botulinum toxin) on the development of neuromuscular correlations was studied in the chick embryo. Embryos chronically injected with toxins showed a marked reduction in the growth and differentiation of skeletal muscles. In the case of α-toxin, moreover, damage was observed also presynaptically, with loss and degeneration of motoneurons and high reduction in the number of neuromuscular synapses. The possibility of the existence of a myogenic signal, which controls in a retrograde manner the growth of the motoneuron and the stability of the synapse, is discussed.

References

1 *Boethius, J.:* The development of electromyogram in the chick embryo. J. exp. Zool. *165:* 419–424 (1967).
2 *Burt, A.M.:* Choline acetyltransferase and neuronal maturation; in *Ford* Neurobiological aspects of maturation and aging. Progress in brain research, vol. 40, pp. 245–252 (Elsevier, Amsterdam 1973).
3 *Cajal, S.R.:* Histologie du système nerveux (Maloine, Paris 1909).
4 *Drachman, D.B.:* Is acetylcholine the trophic neuromuscular transmitter? Archs Neurol., Chicago *17:* 206–218 (1967).
5 *Filogamo, G.:* Neurogenic control versus autonomous determination of muscle cell development; in *Tauc* Synaptogenesis, pp. 31–43 (Naturalia & Biologia, Jouy-en-Josas 1976).
6 *Filogamo, G. and Gabella, G.:* The development of neuromuscular correlations in vertebrates. Archs Biol., Liège *78:* 9–60 (1967).
7 *Filogamo, G. and Marchisio, P.C.:* Acetylcholine system and neural development; in *Ehrenpreis and Solnitzky* Neurosciences research, vol. 4, pp. 29–70 (Academic Press, New York 1970).
8 *Giacobini, G.:* Embryonic and postnatal development of choline acetyltransferase activity in muscles and sciatic nerve of the chick. J. Neurochem. *19:* 1401–1403 (1972).
9 *Giacobini, G.; Filogamo, G.; Weber, M.; Boquet, P., and Changeux, J.-P.:* Effects of a snake α-neurotoxin on the development of innervated skeletal muscles in chick embryo. Proc. natn. Acad. Sci. USA *70:* 1708–1712 (1973).
10 *Giacobini-Robecchi, M.G.; Giacobini, G.; Filogamo, G., and Changeux, J.-P.:* Effects of the type A toxin from *Clostridium botulinum* in the development of skeletal muscles and of their innervation in chick embryo. Brain Res. *83:* 107–121 (1975).
11 *Giacobini-Robecchi, M.G.; Giacobini, G.; Filogamo, G. et Changeux, J.-P.:* Effets comparés de l'injection chronique de toxine α de *Naja nigricollis* et de toxine botu-

linique A sur le développement des racines dorsales et ventrales de la moelle épinière d'embryons de poulet. C.r. hebd. Séanc. Acad. Sci., Paris *283:* 271–274 (1976).

12 *Hebb, C.O.; Krnjevic, K., and Silver, A.:* Acetylcholine and choline acetyltransferase in the diaphragm of the rat. J. Physiol., Lond. *171:* 504–513 (1964).

13 *Hirano, M.:* Ultrastructural study on the morphogenesis of the neuromuscular junction in the skeletal muscle of the chick. Z. Zellforsch. mikrosk. Anat. *79:* 196–208 (1967).

14 *Sisto-Daneo, L. and Filogamo, G.:* Ultrastructure of developing myoneural junctions. Evidence for two patterns of synaptic area differentiation. J. submicrosc. Cytol. *6:* 219–228 (1974).

15 *Visintini, F. e Levi-Montalcini, R.:* Relazioni fra differenziazione strutturale e funzionale dei centri e delle vie nervose nell'embrione di pollo. Schweizer Arch. Neurol. Psychiat. *44:* 119–150 (1939).

G. Giacobini, MD, Department of Human Anatomy, University of Torino, Corso M. D'Azeglio 52, *I–10126 Torino* (Italy)

Properties of Acetylcholine Receptor Turnover in Cultured Embryonic Muscle Cells

John M. Gardner and Douglas M. Fambrough

Carnegie Institution of Washington, Department of Embryology, Baltimore, Md.

Introduction

The turnover rate of the acetylcholine receptor (AChR) in embryonic muscle in tissue culture has been measured by several different methods. The original measurements of the receptor protein half-life were based on the time dependent degradation of ^{125}I-mono-iodo-α-bungarotoxin bound to the receptor (5).

It was reasoned that the toxin receptor interaction was so strong that the complex would be degraded as a unit by the cells and a variety of tests indicated that the binding of α-bungarotoxin did not greatly alter the rate of receptor turnover. These experiments led to the establishment of a 22-hour half-life for chick embryonic toxin receptor complexes under normal culture conditions. However, it remained desirable to obtain a direct measure of the turnover rate of the native receptor. This was accomplished in developing calf myotubes by means of the biosynthetic incorporation of radioactive ^{35}S-1-methionine into newly synthesized receptor protein which was subsequently purified by affinity chromatography (7). This measure of degradation based on pulse labeling of the cells indicated that the receptor protein has a half-life of 17 h, a value which was identical to that obtained indirectly by measurement of the degradation of ^{125}I-α-bungarotoxin bound to the receptor sites of developing calf myotubes in culture.

The purpose of this study was to obtain a direct measure of receptor turnover utilizing still another technique, that is, the incorporation of amino acids labeled with heavy isotopes into newly synthesized receptors and the subsequent kinetic analysis of the disappearance of normal or density shifted receptors. The materials and methods employed in these experiments have been previously described (6).

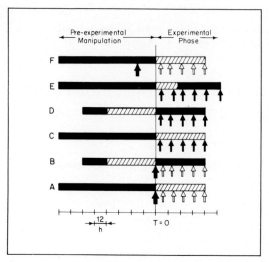

Fig. 1. Experimental strategies employed in ACh receptor turnover studies. A = Density shift-toxin degradation; B = density washout-toxin degradation; C = density shift; D = density washout; E = pulse chase; F = two-population experiment. The solid bars indicate the time intervals during which cells were in normal medium (containing ^1H, ^{12}C, ^{14}N-amino acids). The hatched bars indicate the growth of cells in 'dense' medium (containing ^2H, ^{13}C, ^{15}N-amino acids). The large solid arrows represent the application of ^{125}I-mono-iodo-α-bungarotoxin to all cultures involved in the experiment for a period of time just sufficient to saturate the receptors on the cells, the unbound α-bungarotoxin then being rinsed away and the cultures returned to the incubator. The small solid arrows indicate the time points at which small subsets of cultures were incubated with ^{125}I-α-bungarotoxin, rinsed and then the receptors extracted and analyzed by velocity sedimentation. The small open arrows indicate the same procedure except ^{131}I-α-bungarotoxin was used.

Results

A variety of strategies based on the density shift technique were employed in this study. Each of the experimental designs referred to in the following discussion is depicted graphically and described in detail in figure 1. In order to determine whether cells grown in the presence of dense (^2H, ^{13}C, ^{15}N-labeled) amino acids turned over ACh receptors at the same rate as cells grown in normal medium and whether receptors labeled with dense amino acids turned over at the same rate as those synthesized from normal, light (^1H, ^{12}C, ^{14}N) amino acids, the first two kinds of experiments were performed (fig. 1A, B). These experiments are termed the 'density shift-toxin degradation' experiment and the 'density washout-toxin degradation' experiment, respectively. In the former case myogenic cells were plated out in normal medium and allowed to differentiate and synthesize a set of normal ACh receptors. These receptors were labeled with

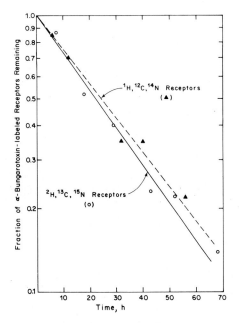

Fig. 2. Degradation of ^{125}I-α-bungarotoxin-receptor complexes involving normal and dense receptors. ○ = Toxin-dense receptor complexes in normal medium; ▲ = toxin-normal receptor complexes in dense medium. Each time point is the average of five small culture dishes. The lines were drawn to fit the data by least squares analysis of data points.

^{125}I-α-bungarotoxin and the cultures switched to medium containing dense amino acids. In the latter case myogenic cells were plated and grown in medium containing dense amino acids so that all of their ACh receptors were dense. These receptors were labeled with ^{125}I-α-bungarotoxin and the cultures were switched back to normal culture medium. In both cases the loss of radioactivity from the cells was monitored for at least 48 h. In the density shift-toxin degradation experiment (fig. 1A), α-bungarotoxin receptor complexes were extracted from the cultures at each time point and analyzed by velocity sedimentation using a ^{131}I-α-bungarotoxin-light-receptor marker in each gradient. In all cases, 100% of the ^{125}I-α-bungarotoxin receptor complexes were 10S, indicating that the labeled α-bungarotoxin never dissociated from the light receptors and rebound to the heavy receptors. Similarly, the ^{125}I-α-bungarotoxin receptor complexes in the density washout-toxin degradation experiments were 100% density shifted throughout the experiment. The results of the two types of experiments are shown in figure 2. In both cases the loss of ^{125}I-α-bungarotoxin receptor complexes followed first order kinetics with a half-time of 22–24 h.

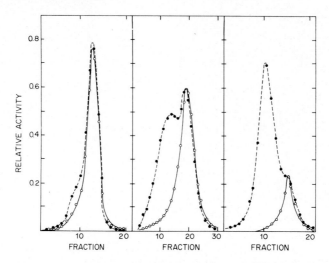

Fig. 3. Sucrose gradient velocity sedimentation profiles of surface receptors extracted from cultures of myogenic cells possessing different proportions of normal and densely labeled receptors after growth in medium containing ^{2}H, ^{13}C, ^{15}N-amino acids. A marker of ^{131}I-mono-iodo-α-bungarotoxin-^{1}H, ^{12}C, ^{14}N receptor complexes (○) was mixed with the sample before centrifugation and its position in the gradient is identical to the position of the ^{125}I-mono-iodo-α-bungarotoxin-^{1}H, ^{12}C, ^{14}N receptor complexes (●).

These values are in good agreement with the established 22-hour half-time of toxin receptor complexes under normal cultured conditions. Thus the turnover of ACh receptors to which α-bungarotoxin is bound takes place at a rate which is not significantly influenced either by culture medium containing ^{2}H, ^{13}C, ^{15}N-labeled amino acids or by a high degree of substitution of these amino acids in the polypeptide chains of the ACh receptors.

In order to measure the turnover rate of the native ACh receptors, without bound α-bungarotoxin, experimental designs C and D of figure 1 were used. These are termed 'density shift' and 'density washout' experiments, respectively. Chick muscle cultures were prepared using normal medium or medium with dense amino acids. After a large population of receptors was present in the cultures, the medium was switched from normal to dense or from dense to normal and cultures maintained in the new medium for the course of the experiment. At each time point, the ACh receptors in a small subset of the cultures were labeled by brief exposure to ^{125}I-α-bungarotoxin and the labeled receptors were solubilized in nonionic detergent and analyzed for normal and density shifted receptors by velocity sedimentation in sucrose gradients. An example of the velocity gradient profiles from cells possessing different proportions of normal light receptors and density labeled receptors is shown in figure 3.

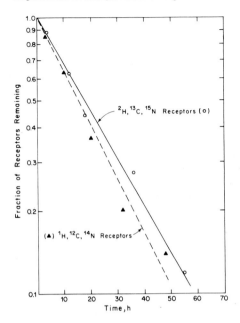

Fig. 4. Degradation of normal and density shifted receptors without bound α-bungarotoxin. ○ = Dense receptors; ▲ = normal receptors.

The total number of normal receptors remaining on the cells switched to dense medium and the total number of density-shifted receptors remaining on the cells switched to normal medium were plotted as a function of time. These curves could readily be approximated by first order exponentials with half-decay times of 15–17 h (fig. 4).

In order to obtain further evidence that unbound ACh receptors turn over at a faster rate than toxin receptor complexes and to confirm that unbound normal light receptors turn over at the same rate as unbound densely labeled receptors the experimental approach E (fig. 1) was adopted. This experiment was termed the 'pulse chase' experiment and was characterized by exposing normal differentiated myotubes to medium containing dense amino acids for a period of 21 h. The time points which were taken during the 'pulse phase' of the experiment reflected the degradation of the population of normal light receptors present at the time the medium was switched to that containing the dense amino acids. For the 'chase phase' of the experiment the cells were placed in normal medium and the disappearance of the population of densely labeled receptors present at the beginning of the chase phase was measured. The rates of degradation of normal receptors, during the pulse phase of the experiment and dense receptors, during the chase phase of the experiment, both fit single exponential

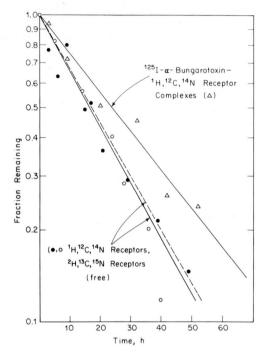

Fig. 5. Degradation of normal and density shifted receptors without bound α-bungarotoxin, employing the pulse chase experimental regime. The open and closed circles represent two experiments (○, ●) combining the degradation data for ^{1}H, ^{12}C, ^{14}N receptors and ^{2}H, ^{13}C, ^{15}N receptors from the pulse and chase phase of the experiments, respectively. The result of one parallel experiment measuring the decay of ^{125}I-α-bungarotoxin receptor complexes from an identically treated set of cultures is also shown (△).

decay curves with half-times of 15–17 h (data not shown). When plotted together, the total numbers of both normal and dense receptors decrease exponentially with a half-life of 16–17 h (fig. 5), a value not different from that of normal and dense receptors, respectively. Furthermore, when a parallel set of cultures was treated with ^{125}I-α-bungarotoxin at the beginning of the chase phase, those toxin receptor complexes turned over at a slower rate, exhibiting a half-time of 23 h as shown in figure 5.

The final experimental design (fig. 1F) was termed the 'two population' experiment. In this experimental design, degradation rates of receptors with and without bound α-bungarotoxin were measured on the same cells. This experiment thus is internally controlled, so any difference in turnover rates cannot be ascribed to variations between experiments. Unfortunately this experimental design is the most difficult, and the results to date only suggest that when α-bungarotoxin is bound to ACh receptors, their degradation rate is somewhat slower than that of normal receptors. Myogenic cells were plated in normal medium and differentiated to produce normal receptors. These receptors were labeled with ^{125}I-α-bungarotoxin and the cultures were returned to normal medium for 18 h, during which time a new set of unlabeled normal ACh receptors was synthesized. Then the cultures were switched to medium containing dense amino acids. The disappearance of the normal receptors during culture in the dense medium was monitored by saturating the free receptors in small subsets of cultures at different times using ^{131}I-α-bungarotoxin and then analyzing the fraction of ^{125}I-α-bungarotoxin-receptor complexes and ^{131}I-α-bungarotoxin-light-receptor complexes after velocity sedimentation of the detergent-extracted receptors. The results from two experiments were as follows: The loss of free normal ACh receptors from the cultures followed first order kinetics with a half-decay time of 18.75 and 20 h. The corresponding ^{125}I-α-bungarotoxin-normal receptor complexes were lost at rates of 19 and 24 h, respectively. Thus receptors to which α-bungarotoxin is bound are degraded slightly slower than free receptors.

Discussion

The details of cholinergic receptor turnover in embryonic muscle cells and adult muscle are important to the mechanisms of formation and maintenance of neuromuscular junctions and of cholinergic synapses between nerve cells. When discussing the biochemical processes relevant to synaptogenesis, it is essential to consider the possible role of receptor metabolism in the stabilization and localization of receptor protein (3).

In these experiments we have measured the degradation rate of ACh receptors, both directly using a number of experimental designs based on the density shift technique and indirectly by measuring the decay of bound α-bungarotoxin. The value of 17 h for the receptor half-life obtained directly by means of the density shift technique applied to chick skeletal myotubes in culture, is in good agreement with the direct measurement previously obtained for developing calf myotubes in culture (7). However, unlike the case of calf myotubes in culture, the data presented here indicated that ^{125}I-α-bungarotoxin-receptor complexes turn over at a rate somewhat slower than unbound ACh receptors; the toxin-receptor complexes turning over with an average half-life of 22 h and the native receptors displaying a 17-hour half-life. The two measures of degradation, the decay of bound α-bungarotoxin and pulse chase labeling of receptors gave identical half-life values of 17 h for calf myotubes (7). The discrepancy between the two measurements of loss of ^{125}I-α-bungarotoxin from the surfaces of developing chick and calf myotubes in culture could possibly reflect differences inherent in the metabolism of receptors in the two cell types; or it is possible that the two experiments presented, measuring the loss of ^{125}I-α-bungarotoxin from the surface of the calf myotubes were of too short a duration to detect a 5-hour difference in the turnover rates of bound and native receptors.

Measurements in this laboratory over the past 3 years have repeatedly confirmed an average value of 22 h for the loss of toxin from the surface of chick skeletal myotubes (4).

In view of these results, the question arises whether toxin binding is a valid probe of receptor metabolism. The discovery of the differential stability of the ACh receptor in extra-junctional and junctional areas of the denervated or embryonic rat diaphragm (1, 2), which was made by measuring the loss of specifically bound toxin, underlines the importance of this line of experimentation. The basic result that extra-junctional receptors have a relatively high turnover rate ($T_{1/2}$ of 8–22 h) and the junctional receptors are more stable ($T_{1/2}$ >100 h) is unlikely to be affected by the data presented here concerning the effect of bound α-bungarotoxin on receptor turnover. However, it appears that future results obtained by application of α-bungarotoxin will have to be interpreted carefully in view of the effect of α-bungarotoxin on turnover rate of the receptor.

There are many possible explanations for the effect of bound α-bungarotoxin on receptor turnover. For examples, the α-bungarotoxin may (a) provide a slight protection from proteolytic attack by sterically hindering the access of proteases to some portions of the receptor or (b) hold the receptor in a conformation which is more metabolically stable than some other normal conformation or (c) block the spontaneous denaturation of receptor molecules which might contribute to the overall degradation rate or (d) change the frequency with which receptors occur in areas of the membrane internalized by the cell for the transport to the lysosomes.

Several years ago we published the results of studies which indicated that α-bungarotoxin did not have a large effect upon the rate of receptor degradation (5). In this experiment we blocked receptor biosynthesis and monitored the degradation of ^{125}I-labeled-α-bungarotoxin bound to the ACh receptors or the disappearance of the receptors in the absence of α-bungarotoxin. In these experiments we found the equivalent of 847 receptors per unit area of cell surface being degraded each hour in the absence of α-bungarotoxin and 810/h in the presence of the α-bungarotoxin. This kind of experiment was technically difficult and we concluded that the difference in rates lay within experimental error and thus that any perturbation of receptor degradation by α-bungarotoxin must be small. As indicated above, our density shift experiments show that the rate of receptor degradation *is* slightly diminished in the presence of bound α-bungarotoxin, so we asked how should this measured effect of α-bungarotoxin manifest itself in the indirect test? The answer is that if the unperturbed degradation of receptors destroyed 847 receptors per unit area of cell surface per hour, then the degradation of receptors with α-bungarotoxin bound to them should occur at a rate of 808 receptors per unit per hour. Thus there is no conflict between the new data presented in this report and the less precise older data as to the effect of α-bungarotoxin on receptor degradation.

Several different kinds of experiments have suggested that some of the ACh receptors are not readily available for interaction with α-bungarotoxin added to the culture medium. We have termed such sites 'hidden receptors'. One interesting aspect of the interaction of α-bungarotoxin with cultured muscle cells is that at 4 °C fewer sites are labeled than at 37 °C. We hypothesized that this difference might be related to the 'hidden sites' being available for interaction slowly with α-bungarotoxin at high temperature but not at low. Our new data on the degradation of receptors with and without bound α-bungarotoxin provides a less mysterious explanation for much of the data related to 'hidden' receptors. In order to measure the slow binding of α-bungarotoxin to receptors of the supposed 'hidden' class at 37. °C, the experiment must be designed to overcome the problem that muscle continues to incorporate new receptors all the time. Therefore we set up experiments to measure the binding kinetics by adding labeled α-bungarotoxin to sets of cultures at different times and then simultaneously washing the unbound toxin from all cultures at the end of the experiments and measuring the counts bound. The assumption was that at this experimental end-point all of the cultures contained the same total number of receptors. The new data on receptor degradation demonstrate that this assumption is not valid. Since α-bungarotoxin stabilized receptors slightly against degradation, the cultures treated for long times in medium containing α-bungarotoxin actually have extra receptors, those spared from degradation due to the bound toxin. This sparing of receptors will alter the total receptor number in a manner which can be quantitatively predicted if one knows the rates of receptor production and degradation over the time interval of the binding experiment. For example, if the cultures have a steady-state level of receptors in the absence of α-bungarotoxin, then adding α-bungarotoxin will result in an increase of about 15% in total receptor number over a 24-hour period. We have recently done several complicated experiments to evaluate this point. These experiments, which cannot be presented in detail here, confirm that α-bungarotoxin stabilization of receptors against degradation can approximately account for the hypothetical slowly saturable 'hidden' sites we postulated earlier. There still remains a set of 'hidden' receptors: i.e. those which cannot be labeled by extracellular α-bungarotoxin and yet are not recently synthesized (6). This population frequently represents about 10–15% of the receptors in cultured chick skeletal muscle cells.

Summary

The methodology of density labeling of proteins by biosynthetic incorporation of ^{2}H, ^{13}C, ^{15}N-amino acids into nascent polypeptide chains permits the direct measurement of the turnover rate of the acetylcholine receptor in cultured chick skeletal muscle. In this study, receptors synthesized in medium containing ^{2}H, ^{13}C, ^{15}N-amino acids were resolved from ^{1}H, ^{12}C, ^{14}N receptors by velocity sedimentation in sucrose-deuterium-oxide gradients and their proportions determined by computer analysis of the gradient profiles.

The kinetics of turnover of the acetylcholine receptor are identical for developing chick muscle fibers grown in medium containing ^{2}H, ^{13}C, ^{15}N-amino acids or ^{1}H, ^{12}C, ^{14}N-amino acids and the high degree of substitution of normal aminoacyl residues by ^{2}H, ^{13}C, ^{15}N-residues does not affect the turnover rate of the denser receptor. The application of a potent, essentially irreversible blocking agent, α-bungarotoxin, slightly increases the half-life of the receptor in the muscle cell plasma membrane from a value of 17 h for the native unbound receptor to 22 h for the α-bungarotoxin-receptor complex.

References

1 Berg, D.K. and Hall, Z.W.: Loss of α-bungarotoxin from junctional and extrajunctional acetylcholine receptors in rat diaphragm muscle *in vivo* and in organ culture. J. Physiol., Lond. *252:* 771–789 (1975).

2 Chang, C.C. and Huang, M.C.: Turnover of junctional and extrajunctional acetylcholine receptors in rat diaphragm. Nature, Lond. *253:* 643–644 (1975).

3 Changeux, J.P. and Danchin, A.: Selective stabilization of developing synapses as a mechanism for the specification of neuronal networks. Nature, Lond. *264:* 705–712 (1976).

4 Devreotes, P.N. and Fambrough, D.M.: Turnover of acetylcholine receptors in skeletal muscle. Cold Spring Harb. Symp. quant. Biol. *40:* 237–251 (1976).

5 Devreotes, P.N. and Fambrough, D.M.: Acetylcholine receptor turnover in membranes of developing muscle fibers. J. Cell Biol. *65:* 335–358 (1975).

6 Devreotes, P.N.; Gardner, J.M., and Fambrough, D.M.: Kinetics of biosynthesis of acetylcholine receptor and subsequent incorporation into plasma membrane of cultured chick skeletal muscle. Cell *10:* 365–373 (1977).

7 Merlie, J.; Changeux, J.P., and Gros, F.: Acetylcholine receptor degradation measured by pulse chase labeling. Nature, Lond. *264:* 74–76 (1976).

J.M. Gardner, Carnegie Institution of Washington, Department of Embryology, 115 West University Parkway, *Baltimore, MD 21210* (USA)

Biosynthesis of Neurotransmitters and Enzymatic Regulation

Maturation of Neurotransmission. Satellite Symp., 6th Meeting Int. Soc. Neurochemistry, Saint-Vincent 1977, pp. 41–64 (Karger, Basel 1978)

Regulation of Neurotransmitter Biosynthesis during Development in the Peripheral Nervous System[1]

Ezio Giacobini

Department of Biobehavioral Sciences, Laboratory of Neuropsychopharmacology, University of Connecticut, Storrs, Conn.

Introduction

Due to more than a century of very active research, the chick embryo represents the best known model of embryologic development (50, 71). Not only the sequence of formation of the autonomic system is known in detail (1, 42, 44, 75, 81) but recent studies (49) have analyzed the pattern and chronology of migration and differentiation of chick embryo sympathoblasts to an extent that no other neuronal subsystem can rival. This is a great advantage for the developmental neurochemists. For now, the process of development can be correlated with a well-understood mature system and the steps which take place in reaching that final state can be intelligently evaluated. Another advantage to utilizing the chick as a model is that the development of the nervous system can be followed very closely from the earliest stages of the embryo without the need for *in utero* experiments and the effect of mother–fetus interactions as in the case of mammals. In addition, experimental manipulations including administration of drugs can be performed at any stage of development with relative ease. These advantages will become evident in the course of the description of the work done in our laboratory during the last few years.

In autonomic ganglia the immature neuron is capable of synthesizing and storing neurotransmitters (norepinephrine, NE, dopamine, DA, acetylcholine, ACh) beginning at the earliest stages of prenatal development (4–7 days of incubation, d.i.) (18, 24, 31, 67).

After the initial ganglionic innervation (6 d.i.) the process is followed by the development of ganglionic synapses (starting 7–9 d.i.) and finally by the establishment of the peripheral field of innervation (6, 7, 18) at 5–16 d.i.

---

[1] Supported by PHS Grant NS-11496 and by grants from the University of Connecticut Research Foundation.

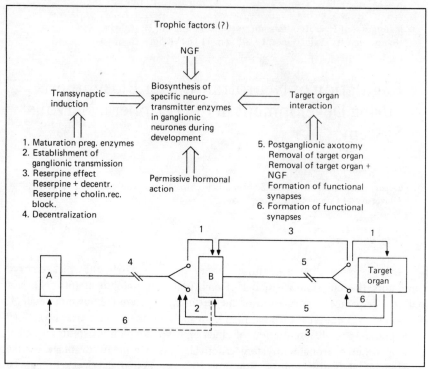

Fig. 1. Mechanisms involved in the regulation of biosynthetic processes during development of the autonomic neurón.

A substantial body of evidence originated from studies in our laboratory (18, 22, 25–27, 29–32, 37, 38, 59) supports the view that the development of enzymes specifically related to neurotransmitter biosynthesis in chick autonomic (sympathetic and parasympathetic) ganglia are regulated by (a) transsynaptic influences, provided by the maturation of preganglionic to postganglionic interactions and (b) retrograde inputs originating in the periphery (target organ) (fig. 1).

Biochemical and pharmacological studies on neuronal control mechanisms of neurotransmitter biosynthesis during the development of the peripheral and central autonomic nervous system, published during the period 1969–1977, have been the object of extensive review by *Giacobini* (30, 31), *Lanier et al.* (48), *Hendry* (41), and *Giacobini and Chiappinelli* (33).

Therefore, in the following sections we shall mainly relate on the most recent results from our own laboratory. The significance of these findings will be discussed in the light of the current literature.

Physiological Development of Neurotransmitter-Related Enzymes in Chick Sympathetic Ganglia

In a series of earlier studies (22, 25, 27, 32, 38, 59) the developmental variations of enzymes associated with the cholinergic synapse and the adrenergic neurons of sympathetic ganglia were investigated in embryonic and posthatching chicken. It was found that the patterns of developmental variations for choline acetyltransferase (ChAc), dopamine-β-hydroxylase (DBH) and monoamine oxidase (MAO) activity were related to (a) the maturation of preganglionic nerve terminals and ganglion neurons and (b) a significant change in the functional state of the ganglion along with an intense activity of the synapses following hatching.

It appeared (22, 38) that during embryonic development the biochemical maturation of ganglionic synapses expressed by ChAc activity correlates well with the development of both the synthesizing (DBH) and the inactivating (MAO) machinery of the adrenergic transmitter in the ganglion neuron.

In a more recent study (25), the developmental variations of tyrosine hydroxylase (TH), the rate-limiting enzyme of catecholamine synthesis and of acetylcholinesterase (AChE) were studied in chick sympathetic ganglia (fig. 3, 4).

In order to relate the biochemical results to the general development of the ganglia, some basic features of ganglionic growth are reported in figure 2. Our period of observation, which lasts from the 5th d.i. to 30 days after hatching (a.h.) is subdivided into three intervals: (1) *ganglionic innervation* (approximately 6–10 d.i.); (2) *cell proliferation and maturation* (12–21 d.i.), and (3) *intense functional activity or posthatching period* (approximately the first 2 weeks a.h.). Sympathetic innervation of peripheral organs (i.e., CA fluorescent fibers) appears in the chick heart at 16 d.i. (24), however, functional innervation is not detected until hatching (68). This demonstrates that the peripheral sympathetic system matures functionally much later than the parasympathetic (cf. functional innervation of the iris at 9 d.i., as described by *Landmesser and Pilar*, (47).

During the period of incubation (fig. 3), the TH activity curve is characterized by a pronounced and steady increase starting from the 12th day of incubation up to day 4 a.h. In the period immediately following hatching (fig. 3) the '4th day fall phenomenon' previously described by us (22, 27) is not seen in the TH curve. Instead, TH activity tends to remain constant between day 2 and 14 a.h. (fig. 3). Both ganglionic protein and weight remain constant during this period (fig. 2), indicating a phase of general pause in protein synthesis. It is interesting to note that both TH and DBH activity which reflect the maturation of the adrenergic neurons peak at day 8 d.i., i.e., just before the peak in NE (43). Following this peak their activity decreases simultaneously to day 12 before

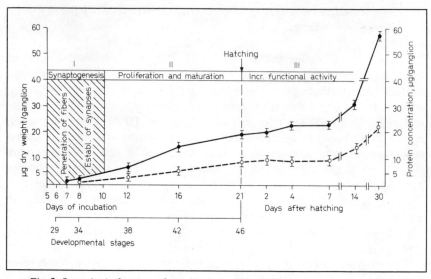

Fig. 2. Some basic features of ganglionic growth at different stages of development of chick sympathetic ganglia. ——— = Dry weight; ---- = protein concentration. From *Fairman et al.* (25).

hatching then shows a parallel pattern of development up to hatching. On the contrary, dopa decarboxylase (DDC) and MAO, which are present in the noradrenergic neuron (27), but are not rate-limiting enzymes, show a different trend increasing steadily from the 6th d.i. up to day 2 a.h. The decrease in TH and DBH activities is simultaneous to the reduction in NE levels and fluorescence in the cell bodies seen by *Jacobowitz et al.* (43) and might express either a slow down in synthesis or an increased transport to the processes.

If we compare the developmental curves of DBH (22) and TH activity, (fig. 3) they show a similar trend before hatching up to day 2 a.h. This corresponds to the period of intense maturation of the sympathetic neuron with an accumulation in the cell body of the two enzymes essential for catecholamine synthesis. The different behavior of TH and DBH in the period following hatching may depend on fundamental differences between the two enzymes such as soluble versus vesicular storage and differences in turnover rate and transport. Our explanation of this first posthatching period 2–7 days a.h. is based on the increased synaptic activity in the chick ganglia following hatching which is demonstrated by the increase in ChAc activity between day 1 and day 3 a.h. (22) (fig. 4). This enhanced synaptic activity may result in the induction of DBH and TH as proposed by *Molinoff et al.* (64, 65), *Black et al.* (8–11), *Thoenen* (76), *Thoenen et al.* (77), and *Black* (3). The inductive response of

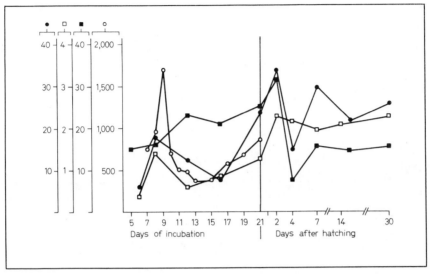

Fig. 3. Developmental pattern of NE, DBH, TH and DDC activity in chicken sympathetic ganglia. ○ = NE, ng/mg protein; ● = DBH, pmol octopamine/μg, protein/h; □ = TH, pmol CO_2/μg dry weight/h; ■ = DDC, pmol CO_2/μg protein/h. Means and standard errors are not reported (for reference see text).

DBH and TH is then followed by the decrease of DBH in cell bodies due to the liberation of this enzyme along with catecholamines at the terminals. Due to the much faster rate of turnover and transport of DBH, as compared to TH, a 'drop' in its activity occurs, instead of a 'plateau' as in TH.

With regard to AChE activity it was found (25) that it increases steadily from the 6th until the 12th d.i. (fig. 4). During this period (fig. 2, period I) neuroblasts actively multiply and undergo differentiation, and synaptic connections are started. In the second half of the embryonic development (fig. 2, period II) and after hatching, AChE remains at a high and constant level (fig. 4). TH and ChAc specific activities actually decrease during the same period, indicating that these enzymes accumulate in ganglion cells at a slower rate, falling behind the general development, while AChE outspeed the general growth of the ganglion. Although these variations might indicate a temporal relationship with the maturation of synapses, it should be considered that AChE in a sympathetic ganglion is present both presynaptically (60%) and postsynaptically (40%) (35) (fig. 7). The increase in nicotinic cholinergic receptors in the embryo, as described by *Greene* (39) seems to be correlated to the development of ChAc activity (fig. 4). However, it occurs several days later (around 13–14 d.i.) and after ChAc activity has already reached a sizable level at 13 d.i. (fig. 4). The developmental curve of the receptor resembles more of the AChE curve and the

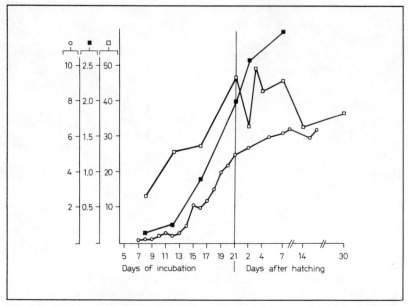

Fig. 4. Developmental pattern of cholinergic receptors, ChAc and AChE activity in chicken sympathetic ganglia. ○ = ACh receptor (α-BTX binding), fmol bound/ganglion; □ = AChE, μmol ACh/mg dry weight/h; ■ = ChAc, nmol ACh/ganglion/h. Means and standard errors are not reported (for reference see text).

binding increases at a slow rate after hatching (fig. 4). It should be kept in mind that a substantial part (60%) of AChE activity and all ChAc activity in sympathetic ganglia is presynaptic (fig. 7), (14, 35). Therefore, preganglionic fibers might influence receptors' appearance and development.

In conclusion. A close temporal relationship in the development of TH and DBH activity is observed throughout the phases of synaptogenesis and maturation but not during the phase of intense functional activity following hatching. Our results strongly suggest that *before hatching* in chick embryo sympathetic ganglia the cholinergic presynaptic terminals play a role in regulating the development of adrenergic postsynaptic enzymes and of the neurotransmitter. In the period *following hatching*, DBH and TH levels in cell bodies are probably regulated by the establishment of the functional activity at the target organs. This results in a depletion of DBH, but not TH, through liberation along with neurotransmitter at the periphery. Reduction of DBH at the terminals might result in increased transport and thereby depletion in the cell body. This mechanism might be responsible for the difference in the pattern of activity of DBH and TH in cell bodies observed in the first week a.h.

Effect of a Single Dose of Reserpine Administered prior to Incubation on Development of TH and ChAc Activity in Chick Sympathetic Ganglia

In this study we examined (26) the effect of a single dose of reserpine administered into the yolk sac prior to incubation on the development of TH and ChAc in sympathetic ganglia of chick embryo and chick.

This represented an attempt to monitor the effect of a transmitter-depleting drug such as reserpine being present from the very beginning of its development in the nervous system and to follow its effect through the first 11 weeks of development. During the first 3 weeks the experiment is performed in a 'closed system' (the egg), i.e. the drug is probably not metabolized (55) to a large extent.

Our results demonstrated that a single dose of as little as 10 μg reserpine (0.18 mg/kg egg) substantially modified the pattern of development of TH activity producing a significant increase at two different periods, the first around the time of hatching and the second at 14–30 days a.h. (fig. 5). Several previous studies (51, 72, 73, 79) demonstrated that reserpine administered prior to incubation is an effective depleter of CA levels in the developing chick brain. According to *Waymire et al.* (79) and *Lydiard and Sparber* (51) CA levels in brain are still depleted by 80–90% at day 18–20 of incubation and by 60% up to 3 days after incubation. In our experiments reserpine caused a total loss of histofluorescence for CA in the ganglia up to hatching.

The depletion of available stores of neurotransmitters at the adrenergic terminals as a consequence of the early reserpine treatment causes *two* types of perturbation on the normal process of development.

First, as soon as the neurotransmitter is synthesized in the cell body and transferred to the terminals it is becoming depleted. Second, a persistent peripheral depletion is probably responsible for (a) a strongly decreased function of the peripheral synapse at hatching and (b) a retrograde signal which, acting transsynaptically, might accelerate the biosynthesis of TH (fig. 1). Both phenomena might concur in triggering a secondary compensatory increase in TH synthesis in the cell body in an attempt to reestablish the peripheral CA level to normal (fig. 1, mech. 5).

We have interpreted the changes in the developmental curve as a temporal (36–48 h) shift of the normal curve of development toward earlier stages (fig. 2) as a result of low levels of CA due to the presence of reserpine at the earliest stages of development. The accelerated developmental schedule is maintained for several weeks a.h. (fig. 5).

According to this interpretation the two periods of significant increase in TH activity seen at and after hatching (fig. 5) are only apparent inductions. Moreover, since no increase in TH activity is found after an acute administration of the drug at day 10–11 a.h. and 20–21 a.h., the second period of increased

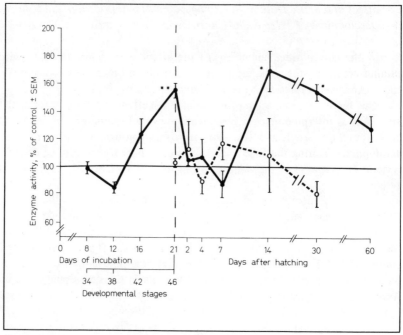

Fig. 5. TH (•) and ChAc (o) activity after reserpine expressed in percent of control in chick sympathetic ganglia. From *Fairman et al.* (26).

TH activity at day 14–30 a.h. (fig. 5) may not be due to the presence of reserpine at that time but to an earlier event such as the long-term depletion of CA.

At a later stage of development, i.e. day 26–27 a.h., the administration of reserpine results in a significant increase of TH activity 72 h later, indicating the presence in the chick of a mechanism of induction similar to that found in rat sympathetic ganglia (6). Our observation is complementary to the results of *Mandell and Morgan* (58) and *Mandell* (56, 57) who were able to induce TH and ChAc activity in both adrenal and brain of 2-week-old chicken by administering 5 mg/kg reserpine twice a day for 3 days.

A premature TH increase triggered by CA depletion could also explain the changes in brain TH activity found by *Lydiard and Sparber* (51) and *Lydiard et al.* (52) in developing chick. Furthermore, as reported below, a similar phenomenon has been observed by *Lysz et al.* (53) for tryptophan hydroxylase in the chick midbrain after serotonin depletion by reserpine.

In sympathetic ganglia of young rat and mice, the transsynaptic regulation of TH development is mediated by the neurotransmitter itself, i.e. ACh, through its action on nicotinic cholinergic receptors, since pharmacological blockade of these receptors prevents normal maturation (2, 4, 40). Following our earlier observation — see previous section and *Dolezalova et al.* (22) — of a large increase of ChAc activity immediately following hatching, we postulated that a blockade of the nicotinic receptor through a daily administration of chlorisondamine during the same period should block the response of TH to reserpine. Since this has proved to be true (26) we can assume that nicotinic stimulation is involved in the TH induction which is apparent at 14 days a.h. (fig. 4, 5). Our results showed that a pharmacological receptor blockade concentrated within a critical period of development, such as the 1st week a.h., is sufficient to inhibit the development of the inductive response to reserpine several days later.

Since reserpine administration did not affect either DDC in sympathetic ganglia or ChAc activity in the ciliary ganglion, its inductive effect can be considered *selective* for adrenergic sympathetic neurons and *specific* for adrenergic enzymes.

The results of this investigation (26) support the view (29–31) that the development of enzyme activities specifically related to neurotransmitter biosynthesis in autonomic ganglia are regulated (a) by transsynaptic influences provided by the maturation of preganglionic to postganglionic interactions and (b) by retrograde inputs originating in the periphery. The effect of reserpine reported here is an example of this second mechanism.

In the *adult organism,* as summarized in the review of *Weiner et al.* (80), the induction of TH as a consequence of factors (stimulation or drugs) which increase sympathetic activity has been demonstrated in peripheral organs and sympathetic ganglia. This mechanism has been recently shown by direct stimulation of the ganglion (82).

Studies on Cholinergic Enzymes in Chick and Pigeon Ciliary Ganglion and Iris Muscle Cells during Synapse Formation

Distribution of Enzymes and Denervation Experiments

The avian ciliary ganglion has been extensively used in the last several years in our laboratory as an alternative peripheral model of development. The ganglion consists of two types of cell bodies which belong to the ciliary and choroid cells (fig. 6). The presynaptic input consists of fibers originating in the accessory oculomotor nucleus (66) (fig. 6). The postsynaptic fibers from the ganglion cells form nicotinic junctions with the striated muscle of the eyes and ciliary muscle (fig. 6), while the choroid fibers innervate the smooth muscle choroidal coat of the eye via muscarinic receptors (45) (fig. 6). The developmental morphology of the ciliary ganglion has recently been studied in detail by *Cantino and Mugnaini* (17) in our Department.

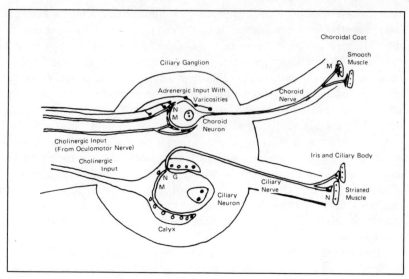

Fig. 6. Diagram of the chick ciliary ganglion and iris.

In a series of experiments, *Martin and Pilar* (60–62) demonstrated that synaptic transmission through the ganglion is accomplished by electrical as well as chemical mechanisms. The ciliary cells combine both electrical and chemical transmission, while choroid cells are activated by purely chemical transmission (63). The development of the gap junctions which are responsible for the electrical coupling have been described by *Cantino and Mugnaini* (17). For our purpose it should be remembered that both choroidal and ciliary cells receive a cholinergic input and form cholinergic synapses with their respective end-organs; therefore, neurochemically speaking the two groups are similar. Pharmacologically it has been demonstrated that choroid cells show a greater sensitivity to blockade by hexamethonium, while the ciliary cells are most sensitive to *d*-tubocurarine (63). A network of adrenergic fibers has been recently described in the avian ciliary ganglion by *Ehinger* (23) and *Cantino and Mugnaini* (16). These fibers apparently do not form synaptic connections in the ganglion (fig. 6). Although, from a biochemical point of view, the fact that both pre- and postsynaptic elements are cholinergic can be regarded as a disadvantage, the ciliary ganglion can be regarded as unique insofar as in the same preparation, the ganglionic elements, their axons (ciliary and choroid nerves) as well as the end-organ of the ganglion, the iris, can be examined.

Since the predominant neural element in the iris is derived from the ciliary cells, the ChAc activity in the target organ can be expected to reflect changes of the activity in the ciliary cell bodies.

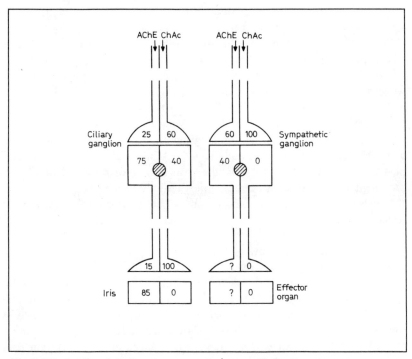

Fig. 7. Distribution of cholinergic enzymes in autonomic ganglia and target organs.

In order to elucidate the distribution of the enzymes *Suszkiw et al.* (74) and *Giacobini et al.* (36) examined the effect of preganglionic denervation and axotomy on the enzymes of ACh metabolism in the pigeon ciliary ganglion cells and iris. The results can be summarized as follows (78):

(1) Preganglionic denervation of ciliary ganglia is followed by a 58% reduction of ganglionic ChAc activity within the first 3 days. Since synaptic endings are degenerated and ganglionic transmission is undetectable during this period, the decrease presumably represents the presynaptic portion of the enzyme. 3 days after denervation there is a further decrease in the ChAc activity to about 15% at day 10, representing about 40% of the postsynaptic portion.

(2) With regard to AChE activity, an initial but less pronounced decrease of about 25% occurs at the 4th day, a second phase shows an additional decrease to about 60% at day 12, which is probably due to postsynaptic AChE.

(3) No significant changes in ChAc and AChE activity are observed in the iris during the first 6 days following preganglionic denervation, but a significant decrease of about 30% ($p < 0.001$) occurs starting on day 7 for ChAc and on day 90 for AChE.

(4) In contrast to the variations in ChAc and AChE activity, there are no significant differences in the ganglionic protein content, lactic dehydrogenase (LDH) activity and coenzyme A (CoA) levels between denervated and control ciliary ganglia.

(5) 3 days after axotomy, ChAc activity in the iris reaches a value of 3% while AChE activity is not substantially changed. Axotomy does not exert any early (3 days) effect on perikaryal ChAc activity.

From these results we obtain the following evidence:

(a) ChAc is present in the normal ganglion, 60% in the presynaptic element and 40% in the ganglion cell body (fig. 7). AChE is 25% presynaptic and 75% postsynaptic (fig. 7). In the iris the distribution of ChAc is at least 97% presynaptic while AChE is present almost entirely in the muscle (85%) (fig. 7), the effect of preganglionic denervation is not confined to the degenerating terminals of the ganglion cells but is conveyed to the postsynaptic elements, i.e. the effect is *transsynaptic.*

(b) The axoplasmic flow of ChAc and AChE is related to the perikaryal concentration of these enzymes.

It appears, therefore, that a transsynaptic regulation of ChAc and AChE synthesis in the ganglion and an intact innervation of the ganglion cells is required in order to maintain normal levels of these enzymes as well as of ACh (69) in the cell bodies. The activities of the nonspecific enzymes LDH and CoA as well as total protein content, remain unchanged after denervation lending support to the theory that the transsynaptic effect regulating the snythesis of ChAc and AChE in the adult ganglion represents a *specific* mechanism.

Developmental Study

In our developmental study (18) we measured ChAc and AChE activity in chick ciliary ganglia and irises from the 5th d.i. until 1 week a.h. The changes in enzyme activity were correlated in time with previous electrophysiological and morphological findings of synapse formation in the same structures.

Low levels of ChAc activity were measured at stage (ST) 26 in the iris (fig. 8a), which implies that as soon as the migration of the ganglion cell is completed (4 d.i., ST 25) the ganglion cells are biochemically differentiated and ChAc is transported down their axons as they grow into the primitive iris (4–7 d.i.). The specific activity of ChAc and AChE rises during development and the increase is closely correlated to the onset and maturation of ganglionic (5–8 d.i.) and iris synaptic transmission (9–10 d.i.).

Two sets of data seem to be most significant:

(1) The 200-fold specific increase of ChAc in iris nerve terminals which occurs at ST 34 (fig. 8a) probably reflects an increase in synthesis of the enzyme in ganglionic cell bodies and suggests that the formation of the iris neuromuscular junction triggers the enzyme induction. This implies the fact that the cell

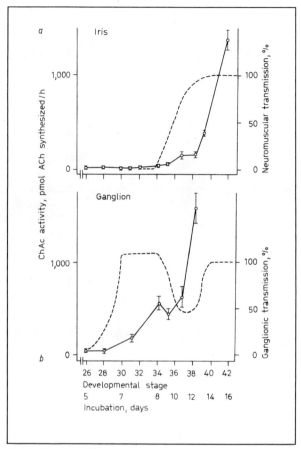

Fig. 8. Relationship between ChAc activity and percent transmission in ciliary ganglion and iris. From *Chiappinelli et al.* (18).

responds to a signal *retrogradely* ascending the axon from the terminals (fig. 1, mech. 5).

(2) The initial increase of AChE-specific activity in the ganglion occurs soon after transmission is established in all cells between ST 30 and 34 and is probably mainly due to enzyme synthesis by the ganglion cells (fig. 9a). This early increase of AChE specific activity in ganglia is chronologically correlated to synapse formation. Since the time courses of specific induction for AChE and ChAc in ganglia are different, it seems clear that the synthesis of these enzymes is regulated by separate mechanisms. In the iris there is a 2-fold increase in the specific activity after the formation of neuromuscular junction, which probably

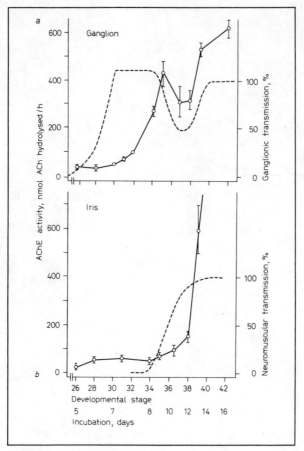

Fig. 9. Relationship between AChE activity and percent transmission in ciliary ganglion and iris. From *Chiappinelli et al.* (18).

reflects enzyme induction in the muscle subneural region (fig. 9b). *We conclude that the specific induction of the AChE in postjunctional cells is due to an influence of the prejunctional elements.*

The growth of ChAc activity is exponential in both ganglia and iris after ST 32 except the period which coincides with the cell death observed by *Landmesser and Pilar* (46) (fig. 8). It is also interesting to note that only very low levels of ChAc are present in both the ganglion and iris when 100% transmission is already present. This demonstrates that the greatest induction of ChAc occurs after functional connection has been established in the tissues. Since the initial increase in ChAc seen in the ganglion between ST 26 and 34 is not present in the iris (fig. 8) during these stages, it suggests that presynaptic elements are responsible for this increase and that formation of functional synapses is closely followed in time by increased ChAc activity in presynaptic terminals.

Our results clearly show that during the normal development of the ganglion, the formation of functional junctions in the target organs originates a *retrograde signal* which reaches the ganglion cell bodies and triggers a rapid (24–36 h) and large-scale synthesis of enzymes needed to sustain the neurotransmitter level at the axon terminal.

A similar event may take place in the presynaptic neurons of the ganglion since ChAc activity in the presynaptic terminal is very low until functional innervation occurs. Thus, in both the ganglion and the iris, ChAc activitiy seems to be induced in the presynaptic elements soon after functional synapses are formed.

The nature of the signal which leads to the induction of ChAc in the presynaptic elements and of AChE in the postsynaptic cells after functional innervation is established is not known. However, the same signal may be responsible for the maintenance of enzyme synthesis in the adult, as demonstrated by the denervation experiments (36, 74). This signal might be the release of ACh and the following depolarization in the postsynaptic cell. Such a role for ACh and for the subsequent cholinergic receptor activation has been suggested during postnatal development in mouse and rat superior cervical ganglia (5, 77) and in chick sympathetic ganglia (22, 25) but not at the neuromuscular junction. An alternative could be the release of a specific messenger from a specific site of the activated membrane.

Our study represents the first direct evidence linking a physiological event, i.e. the establishment of functional neurotransmission, with the regulation of the neurotransmitter enzymes in cell bodies and target organs (fig. 1) within a specific period of development.

By examining these results it appears that the *morphology, physiology,* and *biochemistry* of this model has been systematically integrated to an extent that few other neuronal systems can rival.

Comparison of Maturation between Sympathetic and Parasympathetic Ganglia

As pointed out earlier in this paper the peripheral sympathetic system matures functionally much later than the ciliary ganglion/iris system. This fact is also evident biochemically by comparing the maturation of the respective enzymes and receptors, i.e. the ACh system in the ciliary ganglion with the NE system in the sympathetic ganglia. The former reaches biochemical maturation in the period 11—21 d.i. and the latter in the period 21 d.i. to 7 days a.h. (fig. 3, 4, 8, 9). These results support our view of an important peripheral role of the target organs on ganglionic maturation.

Role of Ganglionic and Peripheral Receptors in Development

At the neuromuscular junction the nicotinic cholinergic receptor appears on the muscle prior to the development of cholinergic innervation (21, 28). Similarly, the development of the muscarinic receptor in whole brain, studied by means of the binding to ^{3}H-QNB (20) parallels, but precedes that of ChAc and in all regions examined achieves adult concentration before that of ChAc.

The rate of appearance of α-bungarotoxin (α-BTX) binding sites has been examined in developing ciliary ganglia and in iris by *Chiappinelli and Giacobini* (19). Specific α-BTX binding (total ^{125}I-α-BTX bound minus ^{125}I-α-BTX bound in the presence of 1,000-fold excess unlabelled α-BTX) increases 4-fold from 7 to 11 d.i. (from 4.2 to 19 fmol/ganglion), after which the amount of binding remains unchanged up to 4 months (a.h.), the oldest age tested. Specific binding in the ciliary ganglion is inhibited up to 90% by $10^{-5}M$ d-tubocurarine, suggesting that, as at the neuromuscular junction, α-BTX binding is associated with the nicotinic cholinergic receptor.

The large increase in binding of α-BTX in the ciliary ganglion occurs after the initiation of ganglionic transmission, similar to the increases seen in ChAc and AChE activities (18), but while a second period of large increase in ChAc activity begins at 12 d.i., the level of α-BTX binding in the ganglion has already reached its adult value by this time.

In contrast to binding in the ganglion, α-BTX binding in the iris remains very low until 12 d.i., then, soon after functional innervation is established, it increases steadily until 4 months a.h. The total increase from 12 d.i. to 4 months a.h. is almost 40-fold (from 2.8 to 106 fmol/iris).

It is interesting to observe that the highest rate of appearance of receptors in the ganglion is posterior to the onset of ganglionic transmission but simultaneous to the initial rise in both AChE and ChAc activity (7—9 d.i.). This implies that the postsynaptic system interacting with the neurotransmitter (i.e. ACh receptor and AChE) starts developing already before the presynaptic terminals are capable of synthesizing large amounts of ACh. The sequence of synaptic events seems to be the following: (1) appearance of a low number of receptors and release of low levels of ACh; (2) onset of ganglionic neurotransmission; (3) rise first in AChE (postsynaptically) and then in ChAc (presynaptically), and (4) increase in receptor density and ACh synthesis.

Before the 8 d.i. the relationship between number of receptors and ChAc activity is similar in sympathetic and ciliary ganglia, indicating an influence of the preganglionic input. At later stages no direct correlation seems to exist between the two developmental patterns.

In developing chick sympathetic ganglia, blockade of postsynaptic nicotinic receptors abolishes the reserpine-induced increase of TH (26). In the ciliary ganglion and iris the development of ChAc and AChE seems to be regulated by signals arising soon after the functional innervation of these organs is accomplished (18).

Effect of Receptor Blockade prior to Innervation

Based on these observations, the effect of receptor blockade *prior to or during* the critical period of innervation of the ganglia and target organs (iris) was studied in the two systems using selective receptor blockers or receptor binders (19).

In order to blockade nicotinic receptors, at the iris neuromuscular junction α-BTX was used; at the ganglionic synapse chlorisondamine was used (19, 34). It is known that α-BTX does bind to sympathetic ganglia (39), however, it does not block neurotransmission (13, 54). To blockade α- or β-receptors (or both) propranolol and phenoxybenzamine were used (34).

A total of 300 μg α-BTX was injected in the yolk-sac at 4 subsequent days (19). The toxin showed a general trophic effect on the iris reducing its weight by approximately 25%; however, the weights of the ganglion and of the embryo were not affected. In the iris, ChAc and AChE activities, both total and specific, were also significantly reduced.

Since we have previously demonstrated (18) (fig. 7) that ChAc in the avian iris is selectively localized to the nerve terminals, we conclude that the toxin has had an effect on the development of the nerve terminals, as well as on (a) the general development of the organ and (b) AChE activity of the iris muscle. Ganglionic ChAc activity was also reduced. By contrast, AChE activity in the ciliary ganglion, which is predominantly postsynaptic (i.e. in the cell bodies), was *not* affected. Although it is difficult to assign the reduction of ChAc to pre- or postsynaptic structures, the significant reduction of activity at the nerve terminals indicate that, at least partially, this must be postsynaptic. In control experiments on a skeletal muscle (biceps femoris) the effect of the toxin was found to be similar to that on the iris, suggesting a similarity in the receptor properties and its developmental role (see also the chapter by G. *Giacobini,* pp. 23–60).

Following chlorisondamine administration, weight, ChAc and AChE were *all* significantly reduced in the ciliary ganglion at *day 14* of incubation. This suggests that the drug through its blocking effect on nicotinic cholinergic receptors, affects the development of the ganglion, both *pre-* (ChAc) and perhaps *post-* (AChE) synaptically. ChAc activity is also reduced in the iris, however, iris weight and AChE activity are normal. This would suggest that, as expected, because of the different properties of the receptors, the muscular junction is not affected by the drug. The reduced ChAc activity in the iris probably reflects a decreased level of the enzyme in the ganglion cells.

At *day 20* the iris has totally recovered its ChAc activity, while the ciliary ganglion still shows reduced weight, and AChE and ChAc activity. This would suggest a long-lasting effect of the drug or a permanent alteration in the ganglion.

In *sympathetic ganglia* α-BTX, administered under the same conditions, does not show any apparent effect on either ChAc or TH activity at day 14 of incubation. This is in contrast with the effect of chlorisondamine, which causes a significant decrease in weight of the ganglia at day 14 of incubation. This effect is still present at day 20 and is associated with a significant decrease of ChAc activity. At day 14 TH activity is only slightly affected but at day 20 it is significantly increased.

These effects might suggest the possibility that although α–BTX is binding to the ganglionic receptors, as has been demonstrated by *Greene* (39), it is not efficiently blocking transmission through the ganglia, as has been shown by *Magazanik et al.* (54), *Bursztajn and Gersohn* (15), *Brown et al.* (13), and *Brown and Fumagalli* (12).

Our results demonstrate that by using two different agents (such as α-BTX and chlorisondamine) selectively binding to or blocking cholinergic receptors, it is possible to study the developmental role of ganglionic and target organ receptors separately. Chlorisondamine by blocking ganglionic receptors seems to exert *two* developmental effects, first *presynaptically* on ChAc and *subsequently postsynaptically* on TH.

Effect of α- and β-Adrenergic Blockade

Single injections of propranolol or phenoxybenzamine separately injected before incubation resulted in a significant increase of TH activity at 8 d.i. (+ 35%) and 7 days (+ 37%) a.h., respectively (34). Neither drug showed an effect similar to that of reserpine reported above. However, the injection of a single dose of propranolol (10 mg/kg eggs) associated to phenoxybenzamine (27.0 mg/kg eggs) prior to incubation produces a significant increase (+ 35%) in TH specific activity of sympathetic ganglia at 20 d.i. (fig. 10) (34). The weight of the ganglion is strongly reduced. This effect is similar to the one seen after chlorisondamine at the same age. At age 14 days a.h. the effect is reversed and both total and specific TH activity are significantly decreased. It is possible that a long-lasting blockade of both α- and β-receptors produces a gradual reduction in preganglionic sympathetic activity leading to reduced enzyme activity. A

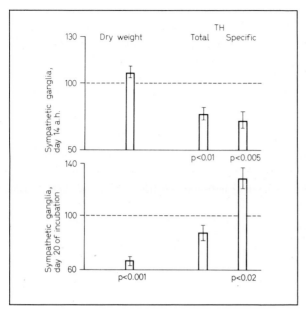

Fig. 10. Effect of a single injection of propranolol and phenoxybenzamine prior to incubation. From *Giacobini and Fairman,* to be published.

similar mechanism has been suggested by *Raine and Chubb* (70) to explain the reduction of TH and DBH in rabbit sympathetic ganglia after long-term adrenergic blockade and the hypotensive effect of propranolol.

Conclusion

Although evidence obtained by using two specific receptor binders confirms our previous view that functional innervation is correlated to the development of both ciliary ganglia and iris, the effect of peripheral receptor blockade at the target organ (iris) is not comparable to the removal of the target organ (fig. 1, mech. 5) as performed by other authors (47).

In the presence of α-BTX binding peripheral junctions can still develop, however, both general effects on weight and specific effects on enzymes are observed. These effects are both pre- and postsynaptic and are similar in the skeletal muscle and in the iris. The effect on ciliary cells is less pronounced, indicating that normal innervation of the iris is not the only signal which allows the survival of ganglion cells. Moreover, blockade of nicotinic receptors on ganglionic (ciliary and sympathetic) cells exerts a general hypotrophic effect on ganglionic development involving both weight and enzymes. However, in sympathetic ganglia this hypotrophic effect on transmitter enzymes is not apparent, indicating that the regulatory signal is not specific. This confirms the findings of other authors (2, 4, 5) in developing mammalian sympathetic ganglia. Blockade of ganglionic receptors in ciliary ganglia prior to innervation does not affect target organ synapses (in the iris). In other words, if normal transmission in ciliary ganglia is unpaired, a situation mimicked by the chlorisondamine experiment, the development of the ganglion is retarded, while the target organ (iris) develops normally. In this case, the blockade of neurotransmission seems to exert a 'local' effect involving only the structures immediately surrounding the blocked junction, such as nerve terminals and postsynaptic membrane components (fig. 1, mech. 6). Our previous results (31) demonstrating specific interactions (acting in both directions) between ganglionic cells and target organs are confirmed by this study (fig. 1). In order to accomplish their regulatory effect on the normal development of synapses, functional receptors are required at the target organs. However, their role seems to be less 'specific' with regard to the development of synapses in ganglia than at the periphery.

Summary

Regulatory mechanisms involved in the development of neurotransmitter biosynthesis are analyzed by using two different biological models in the chick such as sympathetic and parasympathetic (ciliary) ganglia and their target organs. The study includes activities of enzymes related to neurotransmitters, neurotransmitter levels and receptors. A comparison of biochemical as well as neurophysiological and morphological data is attempted. Studies on the development of autonomic neurons and synapses under various pharmacological conditions are reported such as prenatal administration of (a) reserpine, (b) receptor blockers and (c) neurotoxins. The results emphasize the importance of target organ regulation on ganglionic development and the regulatory effect of receptors on the maturation of synapses in ganglia and peripheral organs.

References

1 *Andrew, A.:* The origin of intramural ganglia. IV. The origin of enteric ganglia, a critical review and discussion of the present state of the problem. J. Anat. *108:* 169–184 (1971).
2 *Black, I.B.:* Development of adrenergic neuron *in vivo*. Inhibition by ganglionic blockade. J. Neurochem. *20:* 1265–1267 (1973).
3 *Black, I.B.:* Growth and development of cholinergic and adrenergic neurons in a sympathetic ganglion: reciprocal regulation at the synapse; in Dynamics of degeneration and growth in neurons, pp. 455–467 (Pergamon Press, Oxford 1974).
4 *Black, I.B. and Geen, S.C.:* Transsynaptic regulation of adrenergic neuron development. Inhibition of ganglionic blockade. Brain Res. *63:* 291–302 (1973).
5 *Black, I.B. and Geen, S.C.:* Inhibition of the biochemical and morphological maturation of adrenergic neurons by nicotinic receptor blockade. J. Neurochem. *22:* 301–306 (1974).

6   *Black, I.B. and Reis, D.J.:* Ontogeny of the induction of tyrosine hydroxylase by reserpine in the superior cervical ganglion, nucleus locus coeruleus and adrenal gland. Brain Res. *84:* 264–278 (1975).
7   *Black, I.B. and Mytilineou, C.:* Trans-synaptic regulation of the development of end organ innervation by sympathetic neurons. Brain Res. *101:* 503–521 (1976).
8   *Black, I.B.; Hendry, I., and Iversen, L.L.:* Transsynaptic regulation of growth and development of adrenergic neurons in mouse sympathetic ganglia. Brain Res. *34:* 229–240 (1971).
9   *Black, I.B.; Hendry, I.A., and Iversen, L.L.:* Regulation of the development of choline acetyltransferase in presynaptic nerves by postsynaptic neurons in mouse sympathetic ganglia. J. Physiol., Lond. *216:* 41–42 (1971).
10 *Black, I.B.; Hendry, I.A., and Iversen, L.L.:* Effects of surgical decentralization and nerve growth factor on the maturation of adrenergic neurons in a mouse sympathetic ganglion. J. Neurochem. *19:* 1367–1377 (1972).
11 *Black, I.B.; Hendry, I.A., and Iversen, L.L.:* The role of postsynaptic neurones in the biochemical maturation of presynaptic cholinergic nerve terminals in a mouse sympathetic ganglion. J. Physiol., Lond. *221:* 149–159 (1972).
12 *Brown, D.A. and Fumagalli, L.:* Dissociation of α-bungarotoxin binding and receptor block in rat superior cervical ganglion. Brain Res. *129:* 165–168 (1977).
13 *Brown, D.A.; Garthwaite, J.; Hayashi, E., and Yamada, S.:* Action of surugatoxin on nicotinic receptors in the superior cervical ganglion of the rat. Br. J. Pharmacol. *58:* 157–159 (1976).
14 *Buckley, G.; Consolo, S.; Giacobini, E., and Sjöqvist, F.:* Cholineacetylase in innervated and denervated sympathetic ganglia and ganglion cells of the cat. Acta physiol. scand. *71:* 348–356 (1967)
15 *Bursztain, S. and Gersohn, M.D.:* Distinction between the nicotinic receptors of ganglia and neuromuscular junctions by means of snake neurotoxins. Meet. Am. Soc. Neurosci., Toronto 1976.
16 *Cantino, D. and Mugnaini, E.:* Adrenergic innervation of the parasympathetic ciliary ganglion in the chick. Science *185:* 279–281 (1974).
17 *Cantino, D. and Mugnaini, E.:* The structural basis for electronic coupling in the avian ciliary ganglion. A study with thin sectioning and freeze fracturing. J. Neurocytol. *4:* 505–536 (1975).
18 *Chiappinelli, V.; Giacobini, E.; Pilar, G., and Uchimura, H.:* Induction of cholinergic enzymes in chick ciliary ganglion and iris muscle cells during synapse formation. J. Physiol., Lond. *257:* 749–766 (1976).
19 *Chiappinelli, V. and Giacobini, E.:* Rate of appearance of α-bungarotoxin binding sites during development of chick ciliary ganglion and iris. Meet. Am. Soc. Neurosci., Anaheim 1977.
20 *Coyle, J.T. and Yamamura, H.I.:* Neurochemical aspects of the ontogenesis of cholinergic neurons in the rat brain. Brain Res. *118:* 429–440 (1976).
21 *Diamond, J. and Miledi, R.A.:* A study of fetal and newborn rat muscle fibers. J. Physiol., Lond. *162:* 393–408 (1962).
22 *Dolezalova, H.; Giacobini, E.; Giacobini, G.; Rossi, A., and Toschi, G.:* Developmental variations of choline acetyl-transferase, dopamine-β-hydroxylase and monoamine oxidase in chicken embryo and chicken sympathetic ganglia. Brain Res. *73:* 309–320 (1974).
23 *Ehinger, B.:* Adrenergic nerves in the avian eye and ciliary ganglion. Z. Zellforsch. mikrosk. Anat. *82:* 577–588 (1967).
24 *Enemar, A.; Falck, B., and Håkanson, R.:* Observations on the appearance of norepi-

nephrine in the sympathetic nervous system of the chick embryo. Devl Biol. *11:* 268–283 (1965).

25 *Fairman, K.; Giacobini, E., and Chiappinelli, V.:* Developmental variations of tyrosine hydroxylase and acetylcholinesterase in embryonic and posthatching chicken sympathetic ganglia. Brain Res. *102:* 301–312 (1976).

26 *Fairman, K.; Chiappinelli, V.; Giacobini, E., and Yurkewicz, L.:* The effect of a single dose of reserpine administered prior to incubation on the development of tyrosine hydroxylase activity in chick sympathetic ganglia. Brain Res. *122:* 503–512 (1977).

27 *Filogamo, G.; Giacobini, E.; Giacobini, G., and Noré, B.:* Developmental changes of dopa-decarboxylase in chick embryo spinal and sympathetic ganglia. J. Neurochem. *18:* 1589–1591 (1971).

28 *Fishbach, G.D. and Cohen, S.A.:* The distribution of acetylcholine sensitivity over uninnervated and innervated muscle fibers grown in cell cultures. Devl Biol. *31:* 147–162 (1973).

29 *Giacobini, E.:* Biochemistry of synaptic plasticity studied in single neurons; in Biochemistry of simple neuronal models, vol. 2, pp. 9–64 (Raven Press, New York 1970).

30 *Giacobini, E.:* Biochemistry of the developing autonomic neuron; in Advances in experimental medicine and biology, pp. 145–155 (Plenum Press, New York 1971).

31 *Giacobini, E.:* Neuronal control of neurotransmitter biosynthesis during development. J. neurosci. Res. *1:* 315–331 (1975).

32 *Giacobini, E. and Noré, B.:* Dopa decarboxylase in autonomic and sensory ganglia of the cat. Acta physiol. scand. *82:* 209–217 (1971).

33 *Giacobini, E. and Chiappinelli, V.:* The ciliary ganglion. A model of cholinergic synaptogenesis; in Synaptogenesis, pp. 89–116 (Naturalia et Biologia, Paris 1977).

34 *Giacobini, E. and Fairman, K.:* To be published (1977).

35 *Giacobini, E.; Palmborg, B., and Sjöqvist, F.:* Cholinesterase activity in innervated and denervated sympathetic ganglion cells of the cat. Acta physiol. scand. *69:* 355–361 (1967).

36 *Giacobini, E.; Uchimura, H.; Pilar, G., and Suszkiw, J.B.:* The effect of denervation and axotomy on the enzymes of acetylcholine metabolism in the ciliary ganglion cells. 5th Ann. Meet. Am. Soc. Neurochem., New Orleans 1974.

37 *Giacobini, E.; Chiappinelli, V., and Fairman, K.:* Role of ganglionic and peripheral receptors in neuronal development. 8th Ann. Meet. Am. Soc. Neurochem., Denver 1977.

38 *Giacobini, G.; Marchisio, P.C.; Giacobini, E., and Koslow, S.-H.:* Developmental changes of cholinesterases and monoamine oxidase in chick embryo spinal and sympathetic ganglia. J. Neurochem. *17:* 1177–1185 (1970).

39 *Greene, L.A.:* Binding of alfabungarotoxin to chick sympathetic ganglia: properties of the receptor and its rate of appearance during development. Brain Res. *11:* 135–145 (1976).

40 *Hendry, I.A.:* Transsynaptic regulation of tyrosine hydroxylase activity in a developing mouse sympathetic ganglion. Effects of nerve growth factor (NGF), NGF-antiserum and pempidine. Brain Res. *56:* 313–320 (1973).

41 *Hendry, I.A.:* Control in the development of the vertebrate sympathetic nervous system; in Reviews of neuroscience, vol. 2, pp. 149–194 (Raven Press, New York 1976).

42 *His, W., jr.:* Über die Entwicklung des Bauchsympathicus beim Hühnchen und Menschen. Arch. Anat. EntwGesch., suppl., pp. 137–170 (1897).

43 *Jacobowitz, D.M.; Greene, L., and Thoa, N.B.:* SIF cells, chapter 16, pp. 215–222 (Eränkö 1976).

44 *Kirby, M.L. and Gilmore, S.A.:* A correlative histofluorescence and light microscopic study of the formation of the sympathetic trunks in chick embryos. Anat. Rec. *186:* 437–450 (1976).
45 *Landmesser, L. and Pilar, G.:* Selective reinnervation of the cell populations in the adult pigeon ciliary ganglion. J. Physiol., Lond. *211:* 203–216 (1970).
46 *Landmesser, L. and Pilar, G.:* Synaptic transmission and cell death during normal ganglionic development. J. Physiol., Lond. *241:* 737–747 (1974).
47 *Landmesser, L. and Pilar, G.:* Synapse formation during embryogenesis on ganglion cells lacking a periphery. J. Physiol., Lond. *241:* 715–736 (1974).
48 *Lanier, L.P.; Dunn, A.J., and Van Hartesveldt, C.:* Development of neurotransmitters and their function in brain. Rev. Neurosci. *2* (1976).
49 *Le Douarin, N.:* Cell migration in early vertebrate development studied in interspecific chimeras; in Embryogenesis in mammals. Ciba Found. Symp. No. 40, pp. 71–101 (Elsevier, Amsterdam 1976).
50 *Lillie, F.R.:* Development of the chick. Brief edition, chapter 1–7; reviewed by *H. Hamilton* (Holt, New York 1952).
51 *Lydiard, R.B. and Sparber, S.B.:* Evidence for a critical period for postnatal elevation of brain tyrosine hydroxylase activity resulting from reserpine administration during embryonic development. J. Pharmac. exp. Ther. *189:* 370–379 (1974).
52 *Lydiard, R.B.; Fossum, L.H., and Sparber, S.B.:* Postnatal elevation of brain tyrosine hydroxylase activity, without concurrent increases in steady-state catecholamine levels, resulting from dl-γ-methylparatyrosine administration during embryonic development. J. Pharmac. exp. Ther. *194:* 27–36 (1975).
53 *Lysz, T.; Sze, P., and Giacobini, E.:* Action of reserpine on the development of brain tryptophan hydroxylase. 7th Ann. Meet. Am. Soc. Neurochem., Vancouver 1976.
54 *Magazanik, L.G.; Ivanov, A.Y., and Lukomskaya, N.Y.:* The effect of snake venom polypeptides on cholinoreceptors in isolated rabbit sympathetic ganglia. Neurofisiol. *6:* 652–656 (1974).
55 *Manara, L. and Giacobini, E.:* Unpublished results (1977).
56 *Mandell, A.J.:* Drug induced alterations in brain biosynthetic enzyme activity. A model for adaptation to the environment by the central nervous system; in Biochemistry of brain behavior, pp. 97–121 (Plenum Press, New York 1970).
57 *Mandell, A.J.:* Amphetamine induced increase in tyrosine hydroxylase. Nature, Lond. *227:* 75–76 (1970).
58 *Mandell, A.J. and Morgan, M.:* Increase in regional brain acetyltransferase activity induced with reserpine. Comm. Behav. Biol. *4:* 247–249 (1969).
59 *Marchisio, P.C. and Consolo, S.:* Developmental changes of choline acetyltransferase (ChAc) activity in chick embryo spinal and sympathetic ganglia. J. Neurochem. *15:* 759 (1968).
60 *Martin, A. and Pilar, G.:* Dual mode of synaptic transmission in the avian ciliary ganglion. J. Physiol., Lond. *168:* 443–463 (1963).
61 *Martin, A. and Pilar, G.:* Transmission through the ciliary ganglion of the chick. J. Physiol., Lond. *168:* 464–475 (1963).
62 *Martin, A. and Pilar, G.:* An analysis of electrical coupling at synapses in the avian ciliary ganglion. J. Physiol., Lond. *171:* 454–475 (1964).
63 *Marwitt, R.; Pilar, G., and Weakly, J.:* Characterization of two cell populations in avian ciliary ganglia. Brain Res. *25:* 317–334 (1971).
64 *Molinoff, P.B.; Brimijoin, S.; Weinshilboum, R., and Axelrod, J.:* Neuronally mediated increase in dopamine-β-hydroxylase activity. Proc. natn. Acad. Sci. USA *66:* 453–458 (1970).

65 *Molinoff, P.B.; Brimijoin, S., and Axelrod, J.:* Induction of dopamine-β-hydroxylase in rat hearts and sympathetic ganglia. J. Pharmac. exp. Ther. *182:* 116–129 (1972).

66 *Narayanan, C.H. and Narayanan, Y.:* An experimental inquiry into the central source of preganglionic fibers to the chick ciliary ganglion. J. comp. Neurol. *1:* 101–109 (1976).

67 *Olson, L. and Seiger, A.:* Early prenatal ontogeny of central monoamine neurons in the rat. Fluorescence histochemical observations. Arch. Anat. EntwGesch. *137:* 301 (1972).

68 *Pappano, A.J. and Löffelholtz, K.:* Ontogenesis of adrenergic and cholinergic neuroeffector transmission in chick embryo heart. J. Pharmac. exp. Ther. *191:* 468–478 (1974).

69 *Pilar, G.; Jenden, G.J., and Campbell, B.:* Distribution of acetylcholine in the normal and denervated pigeon ciliary ganglion. Brain Res. *49:* 245–256 (1973).

70 *Raine, A.E.G. and Chubb, I.W.:* Long term β-adrenergic blockade reduces tyrosine hydroxylase and dopamine β-hydroxylase activities in sympathetic ganglia. Nature, Lond. *267:* 265–267 (1977).

71 *Romanoff, A.L.:* in The avian embryo. Structural and functional development, chapter IV: The nervous system, pp. 211–362 (MacMillan, New York 1960).

72 *Sparber, S.B. and Shideman, F.E.:* Estimation of catecholamines in the brain of embryonic and newly hatched chickens and the effects of reserpine. Devl Psychobiol. *2:* 115–119 (1969).

73 *Sparber, S.B. and Shideman, F.E.:* Elevated catecholamines in thirty-day-old chicken brain after depletion during development. Devl Psychobiol. *3:* 123–129 (1970).

74 *Suszkiw, J.B.; Uchimura, H.; Giacobini, E., and Giacobini, G.:* Effect of preganglionic denervation on enzymes of acetylcholine metabolism in the ciliary ganglion. Meet. Am. Soc. Neurosci., San Diego 1973.

75 *Tello, J.F.:* Sur la formation des chaînes primaire et secondaire du grand sympathique dans l'embryon de poulet. Trab. L. Rech. Biol. Univ. Madrid *23:* 1–28 (1925).

76 *Thoenen, H.:* Comparison between the effect of neuronal activity and nerve growth factor on the enzymes involved in the synthesis of epinephrine. Pharmac. Rev. *24:* 255–267 (1972).

77 *Thoenen, H.; Hendry, I.A.; Stockel, K.; Paravicini, U., and Oesch, F.:* Regulation of enzyme synthesis by neuronal activity and by nerve growth factor; in Dynamics of degeneration and growth in neurons, pp. 315–328 (Pergamon Press, Oxford 1974).

78 *Uchimura, H.; Pilar, G.; Giacobini, E., and Suszkiw, J.B.:* To be published (1977).

79 *Waymire, J.C.; Vernadakis, A., and Weiner, N.:* Studies on the development of tyrosine hydroxylase, monoamine oxidase and aromatic-L-amino acid decarboxylase in several regions of the chick brain; in Drugs and the developing brain, pp. 149–170 (Plenum Press, New York 1974).

80 *Weiner, N.; Glutier, G.; Bjur, R., and Pfeffer, R.O.:* Modification of norepinephrine synthesis in intact tissue by drugs and during short-term adrenergic nerve stimulation. Pharmac. Rev. *24:* 203–221 (1972).

81 *Weston, J.A.:* A radiographic analysis of the migration and localization of trunk neural crest cells in the chick. Devl Biol. *6:* 279–310 (1963).

82 *Zigmond, R.E. and Chalazonitis, A.:* Regulation of ganglionic tyrosine hydroxylase activity by nerve activity. Meet. Am. Soc. Neurochem., Denver 1977.

E. Giacobini, MD, Department of Biobehavioral Sciences, Laboratory of Neuropsychopharmacology, University of Connecticut, *Storrs, CT 06268* (USA)

Ontogeny of an Embryonic Mouse Sympathetic Ganglion *in vivo* and *in vitro*

Ira B. Black and Michael D. Coughlin

Laboratory of Developmental Neurology, Cornell University Medical College, New York, N.Y.

Introduction

Although sympathetic neurons have been employed extensively as models of neuronal development, mechanisms regulating *embryonic* growth and development of mammalian ganglia are undefined. The prenatal ontogeny of the parasympathetic innervation of mouse submandibular gland has recently been described (15, 16). The pattern of axon outgrowth from submandibular ganglion to gland parallels to morphogenesis of the glandular epithelium in time and space, and development *in vitro* reproduces that *in vivo*.

Regulation of the embryonic development of mammalian sympathetic ganglia remains to be defined. A great deal of attention has focused on the *perinatal* maturation of the mouse, rat and guinea pig sympathetic superior cervical ganglion (SCG) *in vivo* (3, 4, 6–9, 19, 20) and in culture (10, 11, 18, 24, 32, 36). The SCG is anatomically discrete, allowing various *in vivo* manipulations, and possesses specific markers by which presynaptic and postsynaptic neural development may be monitored. For example, the activity of tyrosine hydroxylase (T-OH), the rate-limiting enzyme in catecholamine biosynthesis (28), is highly localized to adrenergic neurons in the SCG (5) and may be employed as an index of maturation of these cells (6). Considerable evidence indicates that the normal postnatal maturation of adrenergic neurons is dependent on anterograde transsynaptic regulation (3, 6, 8, 9), and on target organ influences (19, 21). In turn, target organs may regulate adrenergic development through the mediation of nerve growth factor (NGF) (21, 26). Injection of neonates with NGF results in hypertrophy and increased numbers of adrenergic neurons throughout the body (13, 26), while treatment with antiserum to NGF (Anti-NGF) is associated with almost total sympathectomy (13, 26). Injection of pregnant mice with Anti-NGF late in gestation (15–17 days) leads to partial immunosympathectomy of the offspring, while earlier treatments (7–12 days) only affect heart and spleen innervation (22, 23). In tissue culture of postnatal ganglia, added NGF (26), or a related factor produced by supporting cells (38), is required for survival and neurite outgrowth. The regulation of development of the mammalian *embryonic* SCG remains to be defined.

Since the SCG constitutes an important model of mammalian neuronal growth, we have extended our studies to the embryonic ganglion. The SCG has been isolated from mouse embryos, and its morphological and biochemical ontogeny has been defined *in vivo* and *in vitro* from 13 days of gestation to birth. We report that growth requirements, and hence regulatory developmental mechanisms, differ markedly between embryonic and late fetal ganglia.

Materials and Methods

Culture Methods

Ganglia were excised from embryos and fetuses of SWV mice. After cervical dislocation of the pregnant mouse, embryos were removed, transferred to Hanks' balanced salt solution (HBSS), and ganglia were excised under a dissecting microscope. Culture procedures were similar to those described previously (15). Nutrient mixture F12 was supplemented with 10% fetal calf serum, penicillin (50 units/ml), streptomycin (50 μg/ml) and amphotericin B (0.25 μg/ml) ('basal medium').

NGF was prepared as the 25,000 g supernatant fraction of adult male SWV mouse salivary glands homogenized in 20 vol of iced distilled water (37). Medium containing a 10^4 dilution of the supernatant elicited maximal outgrowth from newborn mouse ganglia and from 8-day chick embryo spinal ganglia. Anti-NGF was obtained from Burroughs-Wellcome Co. and used at a concentration of 0.5% (v/v). The NGF-elicited outgrowth from newborn SCG could be completely prevented by addition of Anti-NGF to a final concentration of 0.05% in the medium.

Isolated ganglia were grown on plastic tissue culture dishes. Ganglia from 13-, 14- and 15-day embryos adhered to the dishes with no difficulty; ganglia from 17- and 18-day fetuses and from neonates did not readily adhere, and were attached using fetal calf serum (33). Ganglia from fetuses aged 17 days and older were sectioned in 2 or 3 pieces for culture. Cultures were maintained at 37 °C in an atmosphere of 95% air, 5% CO_2 and nearly 100% relative humidity. Medium was changed every 2 days.

Histology

Cultured ganglia were stained supravitally with a 0.05% (w/v) solution of Janus Green B in HBSS (17). Nerve outgrowth visualized by the staining procedure was confirmed by examination of cultures with phase microscopy.

Biochemistry

Ganglion pairs were transferred to glass conical tubes and excess fluid was removed. Cultured ganglia were removed from the plastic dishes with a 'tygon policeman'. Ganglion pairs in 10 μl of ice-cold distilled water were homogenized and T-OH activity (2) and total portein (29) were determined as described previously (17).

Statistical Analysis

Data were analyzed employing Student's t test. Multiple data were analyzed with the one way analysis of variance and the Newman-Keuls test.

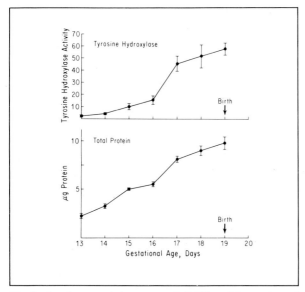

Fig. 1. Developmental increases of T-OH activity and total protein in embryonic ganglia *in vivo*. Groups of 5–11 mice were taken from litters of different gestational ages, and ganglion pairs from each animal were assayed for enzyme activity and total protein (see 'Materials and Methods'). T-OH activity is expressed as mean pmol of product per ganglion·h ± SEM (vertical bars). Total protein is expressed as mean μg per ganglion ± SEM (17).

Results

Growth in vivo

The SCG coalesced to form a well-defined, ovoid tissue mass late during the 13th day of gestation, lying medial to the nodose ganglion, which, in turn, lay medial to the otic cyst. To define normal development of the SCG *in utero*, ganglia from litters of different ages were assayed for T-OH activity and total protein. T-OH activity was detectable in ganglia of 13-day embryos, and increased approximately 100-fold between this time and birth (fig. 1). During the same period, total ganglion protein rose 4-fold, resulting in a marked increase in T-OH specific activity (fig. 1).

Growth in vitro

Tissue culture studies were performed to determine whether development of the embryonic SCG *in vitro* paralleled that *in vivo*, and whether conditions necessary for ganglion differentiation varied during prenatal growth. Ganglion explants from 13-, 14- and 15-day embryos grown in basal medium *without added NGF* exhibited extensive fiber development by 48 h of incubation. Axon

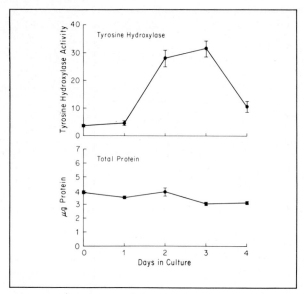

Fig. 2. Development of T-OH activity and total protein in embryonic ganglia in tissue culture. Ganglion pairs were removed from groups of 4–21 mice of gestational age 14 days, and were grown in basal medium for the indicated periods of time. Enzyme activity and total protein are expressed as in figure 1. For T-OH activity the 2- and 3-day values differ from all other times at $p < 0.001$. The total protein values do not differ significantly ($p > 0.05$) (17).

extension was evident after 24 h of culture, continued increasing through 48 h, and, although maintained, exhibited little further development during the third day of incubation. Thereafter, neurites degenerated, while ganglion cells lost cohesiveness and sloughed into the medium.

Cultured ganglia were assayed for T-OH activity to determine whether development *in vitro* paralleled maturation *in vivo*. Explants of 14-day gestation ganglia grown in basal medium without added NGF exhibited a marked increase in T-OH activity during the first 3 days in culture. An initial variable lag period of recovery from the trauma of explantation (30, 33) was reflected in the present studies by a 1-day plateau prior to the increase in enzyme activity. Thereafter, activity increased dramatically, resulting in more than a 8-fold total rise (fig. 2). This *in vitro* increase in T-OH activity paralleled that which occurred *in vivo*. After 3 days in culture T-OH activity was the same as that of the *in vivo* 16- to 17-day ganglion (fig. 1, 2). After the 3rd day in culture, activity began to decline. Explants of 15-day ganglia responded similarly. Although there was no significant increase of total ganglion protein in culture, the values of 4 µg per ganglion approximated that observed *in vivo*, and resulted in a comparable increase in specific enzyme activity (fig. 1, 2).

Table I. Effect of gestational age on growth requirements of ganglia *in vitro*

Gestational age, days	Tyrosine hydroxylase activity				
	zero time controls	basal medium	NGF	Anti-NGF	NGF + Anti-NGF
14	12.6 ± 1.47	33.1 ± 2.22*	103.5 ± 2.71**	37.9 ± 3.08*	33.52 ± 1.69*
18	52.3 ± 4.44	26.3 ± 1.73*	173.7 ± 9.29**	30.75 ± 3.70*	34.8 ± 3.89*

Groups of 5–6 ganglion pairs were removed from embryos at the indicated gestational ages and were incubated for 48 h as described under 'Methods', in the presence of the indicated factors. 'Zero time control' represents activity in unincubated ganglia. T-OH activity is expressed as pmol/ganglion·h (± SEM).
For 14-day groups: * differs from 'zero time' and NGF groups at $p < 0.01$, and does not differ from other groups with single asterisk ($p > 0.05$); ** differs from all other groups at $p < 0.01$.
For 18-day groups: * differs from 'zero time' at $p < 0.02$, and does not differ from other groups with single asterisk ($p > 0.05$); ** differs from all other groups at $p < 0.001$.

Regulatory Mechanisms in vitro

To define the relationship of differentiation in culture to the presence or absence of NGF, 14.5-day ganglia were grown for 48 h in (a) basal medium, (b) basal medium + NGF, (c) basal medium + Anti-NGF or (d) basal medium + NGF + Anti-NGF, and compared to zero time controls. In the *absence* of added NGF, T-OH activity increased approximately 3-fold (table I). (At the time of ganglion explantation, these mice were 12 h older than those represented in figure 2, resulting in a higher 'zero time' value.) Ganglia grown in Anti-NGF or in NGF + Anti-NGF exhibited the same 3-fold increase (table I). Consequently, added NGF was not necessary for normal development of ganglion enzyme activity. Although added NGF was apparently not required for ganglion development, addition of NGF stimulated growth, resulting in an 8.6-fold increase in enzyme activity compared to zero time controls (table I). Neurite development paralleled biochemical differentiation: extensive fiber growth occurred in basal medium without added NGF. Medium with Anti-NGF, and Anti-NGF + NGF supported fiber elaborating similar to that seen in basal medium, whereas addition of NGF elicited a more dense elaboration of fibers (fig. 3). Such NGF-independent development did not occur in later fetal ganglia.

To ascertain whether conditions for differentiation were similar later in gestation, ganglia were removed from 18-day fetuses and grown in the different media described above. In the absence of added NGF, T-OH activity not only failed to increase, but decreased by half (table I). Similar decreases occurred in ganglia cultured with Anti-NGF and NGF + Anti-NGF (table I). Addition of NGF to the medium resulted in over a 3-fold increase in T-OH activity.

Neurite development and biochemical differentiation were again comparable. Virtually no axon outgrowth occurred in the *absence* of added NGF or in media containing Anti-NGF or NGF + Anti-NGF (fig. 4). By contrast, addition of NGF resulted in a dense halo of neurites after 48 h in culture (fig. 4).

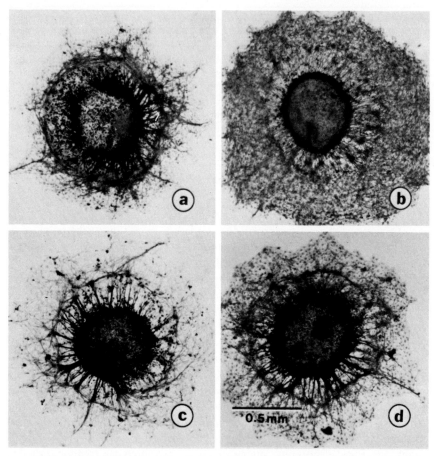

Fig. 3. Embryonic ganglia cultured under different conditions and stained supravitally with Janus Green B. 14-day gestational age ganglion explants were cultured for 48 h under the conditions described in 'Methods'. *a* SCG grown in basal medium without added NGF: Thick bundles of fibers radiate from ganglionic cell mass and terminate in circular neurite meshwork (compare with figure 1). *b* SCG in basal medium + NGF: Dense outgrowth of fibers extends from cell mass. Neuritic production is much greater than in basal medium alone. *c* SCG in basal medium + Anti-NGF: The extent and pattern of neurite outgrowth is essentially the same as in basal medium alone. *d* SCG in basal medium + NGF + Anti-NGF: The extent and pattern of neurite development is basically the same as in basal medium alone. × 35.

Sympathetic Development

Discussion

Our studies were undertaken to define the morphological and biochemical ontogeny of embryonic mammalian sympathetic neurons *in vivo* and in culture.

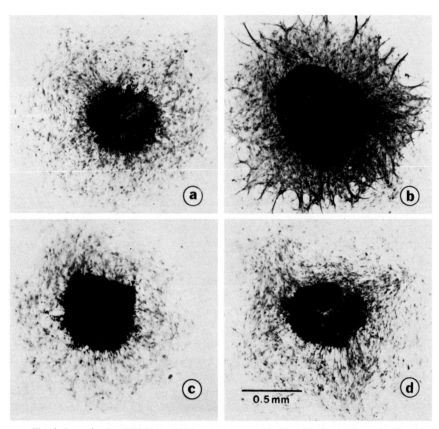

Fig. 4. Late fetal ganglia in culture stained supravitally with Janus Green B. Ganglia were removed from mice on the 18th day of gestation and cultured for 48 h as described in 'Methods'. *a* SCG grown in basal medium without added NGF: A few neurites extend from the cell mass, but axon production is minimal compared to 'b'. *b* SCG grown in basal medium + NGF: Typical dense halo of neurites surrounds cell mass. *c* SCG cultured in basal medium + Anti-NGF. *d* SCG cultured in basal medium + NGF + Anti-NGF. Both 'c' and 'd' exhibit practically no neurite production and are similar to the SCG grown in basal medium alone. Compare to the relatively extensive outgrowth from 14-day gestational age explants in corresponding media (fig. 3). × 35.

The SCG undergoes profound developmental changes well before birth. T-OH activity increases 100-fold *in utero* from the time the ganglion condenses as a discrete mass at 13 days of gestation to birth at 19 days. Total ganglion protein rises 4-fold during this period, resulting in a 25-fold increase in specific T-OH activity. By comparison, T-OH activity in mouse SCG increases only 6- to 8-fold during all of postnatal development (6, 9).

Substantial T-OH activity is present when the ganglion first condenses as a discrete structure. Consequently, this information, encoded within the developing neuron, is initially expressed at or before the neurons arrive at their definitive site(s). Early development of T-OH activity in these mammalian neurons is consistent with reports indicating that the neurotransmitter fate of chick sympathoblasts is determined very early in embryologic life (12, 34).

Tissue culture techniques were employed to analyze regulatory developmental mechanisms. Initial studies indicated that during a specific period of embryologic growth, development *in vitro* corresponded to that *in vivo*. Over a 3-day span the ontogenetic increase of T-OH activity in cultured 14-day ganglia paralleled that *in vivo*. After 3 days in basal medium the embryonic ganglia began degenerating and enzyme activity decreased. In contrast, 18-day ganglia degenerated from culture day zero in basal medium, suggesting that 14-day ganglia cultured for 3 days are comparable to 18-day ganglia *in vivo*, and that growth requirements may change with age. Since cultured ganglia adequately reflected *in vivo* development, experiments were performed to define regulatory mechanisms using *in vitro* methods.

Requirements for growth and development changed radically during embryologic and fetal life as indicated above. Ganglia from 14-day embryos exhibited abundant neurite outgrowth and a 3-fold increase in T-OH activity in the absence of added NGF, or in the presence of Anti-NGF concentrations which entirely inhibited outgrowth from neonatal ganglia. In direct contrast, ganglia from 18-day fetuses exhibited a 50% *decrease* in T-OH activity and essentially no axon elaboration in the absence of added NGF, or in the presence of Anti-NGF, in agreement with previous results obtained with neonatal ganglia (12). However, in both the 14- and 18-day ganglia, addition of NGF to the medium resulted in augmented neurite extension and enzyme activity. It may be inferred that NGF receptors are present in the ganglion before NGF is an absolute requirement for ganglion growth and development.

These observations suggest that there are fundamental ontogenetic differences between embryonic and late fetal ganglia. Our results may indicate that NGF is not an absolute requirement for differentiation of the 14-day ganglion. However, other explanations are possible. In 14-day, but not 18-day ganglia, support cells may produce NGF or an NGF-like substance. This is unlikely, since the neurons developed even in the presence of Anti-NGF, and since in previous reports Anti-NGF has attenuated the putative influence of support cells (38).

Nevertheless, it is conceivable that support cells directly transferred NGF to neurons in an antibody-resistant form in 14-day, but ont 18-day ganglion cultures. Alternatively, different subpopulations of neurons may develop at different times in the ganglion. The 14-day ganglion may contain a subpopulation which develops independent of NGF, as well as one which requires NGF. By 18 days of gestation, the entire ganglion population may require NGF. We are presently attempting to distinguish among the foregoing alternatives. However, we would tentatively favor the view that initial biochemical and morphologic development does not require NGF.

Regardless of the underlying mechanisms, our data suggest that regulatory influences differ markedly in embryonic and late fetal ganglia. Such differences during prenatal development have been defined in nonmammalian systems. Early studies revealed that *chick* sympathetic and sensory ganglia respond to NGF-producing tumors only after the 7th day *in ovo* (27). Existence of a critical period of sensitivity during development has also been documented by experiments *in vitro*. Sympathetic ganglion explants from chick embryos do not respond to NGF at 8 days, become minimally responsive at 9–10 days and exhibit maximum sensitivity at 13–14 days (35). Similar development of sensitivity occurs in chick sensory ganglia: NGF effects are first present at 6 days (25, 39) and become maximal at 7–9 days (27, 39). However, very early (4-day) spinal ganglion neurons do not require added NGF for development in culture (25, 30).

Summary

Development of the embryonic mouse SCG was characterized *in vivo* and in tissue culture. From 13 days of gestation, when the SCG was first visible, to birth at 19 days, T-OH activity increased 100-fold *in vivo*. Explants of ganglia from 14-day embryos cultured for 2 days, exhibited abundant neurite outgrowth in basal medium without added NGF, and increases in T-OH activity paralleled that observed *in vivo*. Ganglia from 14-day embryos elaborated neurites and exhibited 3-fold increases in enzyme activity *in vitro* in the presence of Anti-NGF or NGF + Anti-NGF. In direct contrast, ganglia from 18-day fetuses failed to grow without added NGF, or in medium containing Anti-NGF or NGF + Anti-NGF: virtually no axon outgrowth occurred and T-OH activity decreased by half. These results suggest that developmental regulatory mechanisms changed markedly during embryologic and fetal life of the mammalian SCG.

Acknowledgement

This work was supported by the Dysautonomia Foundation Inc., and the National Science Foundation. *I.B.B.* is the recipient of an Irma T. Hirschl Trust Career Scientist Award. Initial experiments were performed in the laboratory of Dr. *Michel P. Rathbone* under a grant from the Medical Research Council of Canada to the Group in Developmental Neurobiology at McMaster University Medical Centre. SWV mice were a gift of Dr. *Alan Peterson*. We thank Ms. *Elise Grossman* and Ms. *Dahna Boyer* for excellent assistance.

References

1 *Augulis, V. and Sigg, E.B.:* Supravital staining and fixation of brain and spinal cord by intravascular perfusion. Stain Technol. *46:* 183–190 (1971).
2 *Black, I.B.:* Increased tyrosine hydroxylase activity in frontal cortex and cerebellum after reserpine. Brain Res. *95:* 170–176 (1975).
3 *Black, I.B. and Geen, S.C.:* Trans-synaptic regulation of adrenergic neuron development: inhibition by ganglionic blockade. Brain Res. *63:* 291–302 (1973).
4 *Black, I.B. and Mytilineou, C.:* Trans-synaptic regulation of the development of end organ innervation by sympathetic neurons. Brain Res. *101:* 503–521 (1976).
5 *Black, I.B.; Hendry, I.A., and Iversen, L.L.:* Differences in the regulation of tyrosine hydroxylase and dopa decarboxylase in sympathetic ganglia and adrenals. Nature new Biol. *231:* 27–29 (1971).
6 *Black, I.B.; Hendry, I.A., and Iversen, L.L.:* Trans-synaptic regulation of growth and development of adrenergic neurons in a mouse sympathetic ganglion. Brain Res. *34:* 229–240 (1971).
7 *Black, I.B.; Hendry, I.A., and Iversen, L.L.:* The role of post-synaptic neurons in the biochemical maturation of presynaptic cholinergic nerve terminals in a mouse sympathetic ganglion. J. Physiol., Lond. *221:* 149–159 (1972).
8 *Black, I.B.; Hendry, I.A., and Iversen, L.L.:* Effects of surgical decentralization and nerve growth factor on the maturation of adrenergic neurons in a mouse sympathetic ganglion. J. Neurochem. *19:* 1367–1377 (1972).
9 *Black, I.B.; Joh, T.H., and Reis, D.J.:* Accumulation of tyrosine hydroxylase molecules during growth and development of the superior cervical ganglion. Brain Res. *75:* 133–144 (1974).
10 *Bray, D.:* Surface movements during the growth of single explanted neurons. Proc. natn. Acad. Sci. USA *65:* 905–910 (1970).
11 *Chamley, J.H.; Campbell, G.R., and Burnstock, G.:* An analysis of the interactions between sympathetic nerve fibers and smooth muscle cells in tissue culture. Devl Biol. *33:* 344–361 (1973).
12 *Cohen, A.L.:* Expression of sympathetic traits in cells of neural crest origin. J. exp. Zool. *179:* 167–192 (1972).
13 *Cohen, S.:* Purification of a nerve growth-promoting protein from mouse salivary gland and its neurocytotoxic antiserum. Proc. natn. Acad. Sci. USA *46:* 302–311 (1960).
14 *Cohen, S.; Levi-Montalcini, R., and Hamburger, V.:* A nerve growth-stimulating factor isolated from sarcomas 37 and 180. Proc. natn. Acad. Sci. USA *40:* 1014–1018 (1954).
15 *Coughlin, M.D.:* Early development of parasympathetic nerves in the mouse submandibular gland. Devl Biol. *43:* 123–139 (1975).
16 *Coughlin, M.D.:* Target organ stimulation of parasympathetic nerve growth in the developing mouse submandibular gland. Devl Biol. *43:* 140–158 (1975).
17 *Coughlin, M.D.; Boyer, D.M., and Black, I.B.:* Embryologic development of a mouse sympathetic ganglion *in vivo* and *in vitro*. Proc. natn. Acad. Sci. USA *74:* 3438–3442 (1977).
18 *Crain, S.M. and Peterson, E.R.:* Development of neural connections in culture. Ann. N.Y. Acad. Sci. *228:* 6–34 (1974).
19 *Dibner, M.D. and Black, I.B.:* The effect of target organ removal on the development of sympathetic neurons. Brain Res. *103:* 93–102 (1976).
20 *Hendry, I.A. and Iversen, L.L.:* in Proc. 5th Int. Congr. Pharmacol., San Francisco 1972, p. 100.

21 *Hendry, I.A. and Iversen, L.L.:* Changes in tissue and plasma concentrations of nerve growth factor following removal of the submaxillary glands in adult mice and their effects on the sympathetic nervous system. Nature, Lond. *243:* 500–504 (1973).
22 *Klingman, G.I.: In utero* immunosympathectomy of mice. Int. J. Neuropharm. *5:* 163–170 (1966).
23 *Klingman, G.I. and Klingman, J.D.:* Prenatal and postnatal treatment of mice with antiserum to nerve growth factor. Int. J. Neuropharm. *6:* 501–508 (1967).
24 *Ko, C.-P.; Burton, H.; Johnson, M.I., and Bunge, R.P.:* Synaptic transmission between rat superior cervical ganglion neurons in dissociated cell cultures. Brain Res. *117:* 461–485 (1976).
25 *Letourneau, P.C.:* Cell-to-substratum adhesion and guidance of axonal elongation. Devl Biol. *44:* 92–101 (1975).
26 *Levi-Montalcini, R. and Angeletti, P.U.:* Nerve growth factor. Physiol. Rev. *48:* 534–569 (1968).
27 *Levi-Montalcini, R. and Hamburger, V.:* Selective growth stimulating effects of mouse sarcoma on the sensory and sympathetic nervous system of the chick embryo. J. exp. Zool. *116:* 321–362 (1951).
28 *Levitt, M.; Spector, S.; Sjoerdsma, A., and Udenfriend, S.:* Elucidation of the rate-limiting step in norepinephrine biosynthesis in the perfused guinea pig heart. J. Pharmac. exp. Ther. *148:* 1–8 (1965).
29 *Lowry, O.H.; Rosebrough, N.J.; Farr, A.L., and Randall, R.J.:* Protein measurement with Folin phenol reagent. J. biol. Chem. *193:* 265–275 (1951).
30 *Ludueña, M.A.:* Nerve cell differentiation *in vitro.* Devl Biol. *33:* 268–284 (1973).
31 *Mackay, A.:* The long-term regulation of tyrosine hydroxylase activity in cultured sympathetic ganglia: role of ganglionic noradrenaline content. Br. J. Pharmacol. *51:* 509–520 (1974).
32 *Mains, R.E. and Patterson, P.H.:* Primary cultures of dissociated sympathetic neurons. I. Establishment of long-term growth in culture and studies of differentiated properties. J. Cell Biol. *59:* 329–345 (1973).
33 *Mizel, S.B. and Bamburg, J.R.:* Studies on the action of nerve growth factor. I. Characterization of a simplified *in vitro* culture system for dorsal root and sympathetic ganglia. Devl Biol. *49:* 11–19 (1976).
34 *Norr, S.C.: In vitro* analysis of sympathetic neuron differentiation from chick neural crest cells. Devl Biol. *34:* 16–38 (1973).
35 *Partlow, L.M. and Larrabee, M.G.:* Effects of a nerve growth factor, embryo age and metabolic inhibitors on growth of fibres and on synthesis of ribonucleic acid and protein in embryonic sympathetic ganglia. J. Neurochem. *18:* 2101–2119 (1971).
36 *Patterson, P.H. and Chun, L.L.Y.:* The influence of non-neuronal cells on catecholamine and acetylcholine synthesis and accumulation in cultures of dissociated sympathetic neurons. Proc. natn. Acad. Sci. USA *71:* 3607–3610 (1974).
37 *Varon, S.; Nomura, J., and Shooter, E.M.:* The isolation of the mouse nerve growth factor protein in a high molecular weight form. Biochemistry *6:* 2202–2209 (1967).
38 *Varon, S.; Raiborn, C., and Burnham, P.A.:* Implication of a nerve growth factor-like antigen in the support derived by ganglionic neurons from their homologous glia in dissociated cultures. Neurobiology *4:* 317–327 (1974).
39 *Winick, M. and Greenberg, R.E.:* Chemical control of sensory ganglia during a critical period of development. Nature, Lond. *205:* 180–181 (1965).

I.B. Black, MD, Laboratory of Developmental Neurology, Cornell University Medical College, 515 East 71st Street, *New York, NY 10021* (USA)

Role of Nerve Growth Factor for the Development and Maintenance of Function of Sympathetic Neurons and Adrenal Medullary Cells[1]

U. Otten, M. Goedert and H. Thoenen

Biocenter of the University, Basel

Introduction

Nerve growth factor (NGF) is a protein indispensable for the normal development, survival and maintenance of function of peripheral sympathetic neurons (9). A major approach to the elucidation of its physiological role represents the biochemical and morphological analysis of the pleiotypic effects of exogenously administered NGF *in vivo* and *in vitro* and the corresponding characterization of the consequences of the neutralization of endogenous NGF by the administration of monospecific antibodies to NGF.

In this chapter, we will briefly summarize recent studies on the effects of NGF and monospecific purified NGF antibodies on the rat peripheral sympathetic nervous system in the early postnatal phase and after full differentiation. Moreover, we will also include the effects on the adrenal medulla and preliminary results on the consequences of immunization of rats with 2.5 S NGF.

Methods

NGF was prepared as the 2.5 S subunit from submaxillary glands of adult male mice according to *Bocchini and Angeletti* (1). The biological activity was determined by the chick dorsal root ganglion bioassay (3). Antibodies against 2.5 S NGF were raised in goats and purified by affinity chromatography according to *Stöckel et al.* (18).

Tyrosine hydroxylase (TH) activity in superior cervical ganglia and adrenal medullae was assayed according to *Mueller et al.* (14), dopamine β-hydroxylase (DBH) according to *Molinoff et al.* (12), dopa decarboxylase (DDC) according to *Håkanson and Owman* (5) and choline acetyltransferase (CAT) according to *Fonnum* (4). Proteins were determined by the method of *Lowry et al.* (11). All rats were injected subcutaneously (s.c.) with a single dose

---

[1] This work was supported by the Swiss National Foundation for Scientific Research (Grant No. 3.432.74).

of 10 mg/kg of NGF or with 200 mg/kg of purified NGF antibodies. Sprague-Dawley rats (100–150 g) were immunized against NGF by intradermal injection of 400 µg/kg of 2.5 S NGF emulsified in Freund's complete adjuvant and by s.c. injection of pertussis vaccine. Subsequent boosting was performed every 3 weeks with 200 µg/kg of 2.5 S NGF. The presence of NGF antibodies was demonstrated by double immunodiffusion according to *Ouchterlony* (16), the antibody titres were determined by estimating the inhibition of the NGF-mediated fibre outgrowth from chick dorsal root ganglia.

Results

Effects of NGF on Developing and Differentiated Adrenergic Neurons

Beyond the general promotion of growth and the stimulation of fibre outgrowth (9), the selective induction of TH and DBH is one of the most

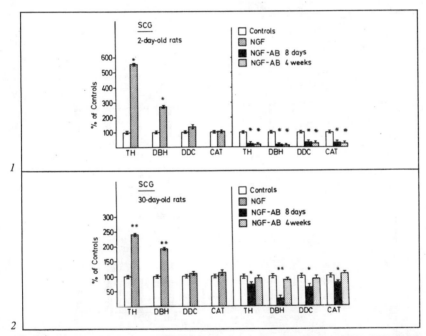

Fig. 1. Effects of NGF and NGF antibodies on the total activities of TH, DBH, DDC and CAT in the SCG of rats injected at 2 days of age. Animals were killed 2 days after NGF, 8 days and 4 weeks after NGF antibody treatment. Each value represents the mean ± SEM (n = 8); * = differs from respective controls at $p < 0.01$.

Fig. 2. Effects of NGF and NGF antibodies on the total activities of TH, DBH, DDC and CAT in the SCG of rats injected at 30 days of age. Animals were killed 2 days after NGF, 8 days and 4 weeks after NGF antibody treatment. Each value represents the mean ± SEM (n = 8); * = differs from respective controls at $p < 0.025$; ** = differs from respective controls at $p < 0.01$.

characteristic effects of NGF on adrenergic neurons (19). This selective induction of two enzymes, which are virtually exclusively localized in adrenergic neurons and adrenal medullary cells (13), can be completely dissociated from the general growth-promoting effect which is reflected by an increase in volume and total protein content of sympathetic ganglia. As shown in figures 1 and 2, both in newborn and 30-day-old animals — at this age the sympathetic nervous system is fully differentiated (20) — the effect of a single dose of NGF is restricted to TH and DBH. The response in the newborn is considerably larger. The effect on the adrenal medulla (fig. 3, 4) which is also independent on the intactness of the preganglionic cholinergic nerves (15) is in principle the same as that of the superior cervical ganglion (SCG), but smaller. The activity of CAT

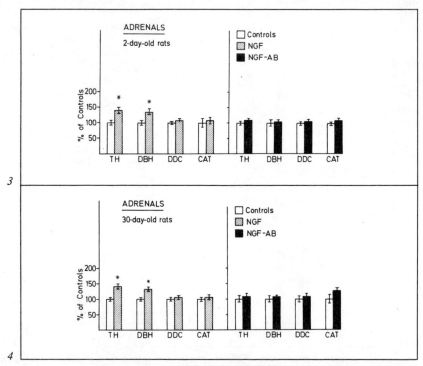

Fig. 3. Effects of NGF and NGF antibodies on the total activities of TH, DBH, DDC and CAT in the adrenals of rats injected at 2 days of age. Animals were killed 2 days after NGF and 8 days after NGF antibody treatment. Each value represents the mean ± SEM (n = 8); * = differs from respective controls at $p < 0.01$.

Fig. 4. Effects of NGF and NGF antibodies on the total activities of TH, DBH, DDC and CAT in the adrenals of rats injected at 30 days of age. Animals were killed 2 days after NGF and 8 days after NGF antibody treatment. Each value represents the mean ± SEM (n = 8); * = differs from respective controls at $p < 0.01$.

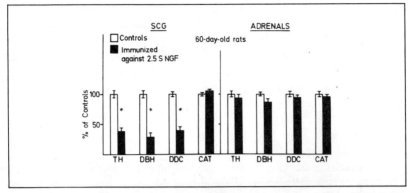

Fig. 5. Effects of immunization with 2.5 S NGF on the total activities of TH, DBH, DDC and CAT in the superior cervical ganglia and adrenals of rats immunized at 20 days of age and boosted twice. The animals were killed 10 days after the second boosting. The NGF antibody titre was 1 mg/ml serum. Each value represents the mean ± SEM (n = 10); * = differs from respective controls at p <0.01.

which is localized to more than 95% in preganglionic cholinergic neurons (6) is not affected in both sympathetic ganglia and adrenal medullae 2 days after a single injection of NGF. This is in contrast to the consistent increase observed in the preganglionic cholinergic nerve terminals of the SCG after repeated administration of NGF (20).

Effects of Monospecific Antibodies to NGF

In agreement with earlier studies the administration of a single dose of NGF antibodies to newborn rats results in a very marked drop of all the enzymes studied (fig. 1). These low levels are reached gradually over several days and the maximal effect is reached after 8 days. The reduction of the enzyme levels is irreversible, reflecting the extensive destruction of adrenergic neurons. In 30-day-old animals the enzyme reduction was smaller (fig. 2) and, most importantly, also completely reversible. Control levels were reached again 4 weeks after NGF antibody administration. In contrast to the newborn animals the reduction of DBH was distinctly larger than that of TH (75 vs. 25%).

Effects of Immunization of Adult Rats with 2.5 S NGF

After previous experiments had shown that the half-life of intravenously injected NGF antibodies was surprisingly short (6 h), we tried a direct immunization with NGF in order to obtain a more continuous level of antibodies to NGF.

Preliminary experiments have shown that 10 days after a second boosting — the antibody titre was 1 mg/ml of serum at that time — the reduction of TH and DBH was considerably larger (fig. 5) than that after a single injection of goat antibodies to NGF (fig. 2). In both cases, the effect on DBH was larger than that on TH. Moreover, this reduction was also accompanied by a decrease in the total protein content of the SCG by 30%. Experiments in progress are designed to establish whether further boosting with NGF will result in an additional reduction of enzyme levels and total protein content and whether these changes are irreversible.

Discussion

The results presented in this chapter support the view that NGF is not only essential for the ontogenetic development of major parts of the peripheral sympathetic nervous system but also for the maintenance of its function after reaching differentiation. This is documented by the observation that the response of adrenergic neurons to exogenously administered NGF is not confined to the early postnatal period but also present in adult animals. Moreover, the neutralization of endogenous NGF or cross-reacting NGF-like molecules by antibodies leads to a distinct impairment of their specific function, namely the synthesis of the adrenergic transmitter norepinephrine as reflected by the reduction of all the enzymes involved in its synthesis. Interestingly, the decrease of DBH is much larger than that of TH and the time-course of the DBH changes also reflects the reduction of the norepinephrine levels in sympathetic ganglia described by *Bjerre et al.* (2). This suggests that under these specific experimental conditions DBH becomes rate-limiting in the synthesis of the adrenergic transmitter.

The response of fully differentiated adrenergic neurons to both NGF and NGF antibodies is smaller than that seen in the early postnatal development. Moreover, the effect of a single injection of NGF antibodies is reversible in adult animals. In very recent experiments which will be published in detail elsewhere, we found that the shift from the irreversible to the reversible effect is a gradual one. Administration of NGF antibodies on day 2 and 6 after birth leads to an irreversible effect. At day 12 the effect is partially and at day 16 fully reversible. The time-course from irreversible to reversible changes is virtually identical to that resulting from the transection of the postganglionic adrenergic nerve fibres (7). Since this effect of axotomy or of destruction of nerve terminals by 6-hydroxydopamine can be prevented by administration of large doses of NGF, it has been concluded that the deleterious effect of axotomy (8) or 6-hydroxydopamine (10) on the cell body in the early postnatal phase results from the interruption of the normal supply of NGF from the periphery. The present experiments strongly support this interpretation, suggesting additionally that the interruption of the axoplasmic transport and the neutralization of endogenous NGF by NGF antibodies produce the same final result. Interestingly, *Purves and*

Njå (17) have shown that exogenous NGF largely prevents the synaptic depression following axotomy in adults.

The fact that in newborn animals NGF antibodies do not lead to a rapid destruction of the adrenergic neurons but to a gradual degeneration and that in adult animals NGF antibodies have only a transient effect, supports the assumption that immunosympathectomy is the result of the deprivation of NGF and does not result from an acute cytotoxic effect. This is further supported by preliminary experiments in organ cultures of the SCG in which the addition of relatively large amounts of NGF antibodies did not reduce the enzyme levels within the first 48 h, even if the system was supplemented with complement. However, a cytotoxic effect was achieved as soon as the antibodies were loaded with NGF. This allows the binding of the antibody via NGF to the cell surface of the adrenergic neuron, making it susceptible to the cytotoxic function of complement. The mediation of the binding of NGF antibodies by NGF is also indicated by the recent observation (*M. Dumas*, unpublished results) that antibodies to NGF are only transported retrogradely if bound to NGF.

The reason why adrenal medullary cells and short adrenergic neurons respond to NGF and not to NGF antibodies is not clear and will be discussed in a forthcoming publication.

Summary

Administration of NGF to newborn and adult rats elicits a selective increase in TH and DBH, both in sympathetic ganglia and adrenal medullae. Monospecific anti-NGF antibodies lead to an irreversible, drastic reduction in TH, DBH, DDC and CAT in sympathetic ganglia of newborn rats. In contrast the reduction in enzymes in adults was smaller and transient. The fact that immunization of adult rats against 2.5 S NGF also reduced all catecholamine-synthesizing enzymes indicates that NGF is not only an absolute prerequisite for the development of sympathetic neurons but that it is also indispensable for maintenance of their function.

References

1 *Bocchini, V. and Angeletti, P.U.:* The nerve growth factor purification as a 30,000 molecular weight protein. Proc. natn. Acad. Sci. USA *64:* 787–794 (1969).
2 *Bjerre, B.; Wiklund, L., and Edwards, C.D.:* A study of the de- and regenerative changes in the sympathetic nervous system of the adult mouse after treatment with the antiserum to nerve growth factor. Brain Res. *92:* 257–278 (1975).
3 *Fenton, E.L.:* Tissue culture assay of nerve growth factor and the specific antiserum. Expl. Cell Res. *59:* 383–392 (1970).
4 *Fonnum, F.:* A rapid radiochemical method for the determination of choline acetyltransferase. J. Neurochem. *24:* 407–409 (1975).
5 *Håkanson, R. and Owman, C.:* Pineal dopa decarboxylase and monoaminoxidase activities as related to the monoamine stores. J. Neurochem. *13:* 597–605 (1966).

6 *Hebb, C.O. and Waites, G.M.H.:* Choline acetylase in antero- and retrograde degeneration of a cholinergic nerve. J. Physiol., Lond. *132:* 667–671 (1956).
7 *Hendry, I.A.:* The effects of axotomy on the development of the rat superior cervical ganglion. Brain Res. *90:* 235–244 (1975).
8 *Hendry, I.A.:* The response of adrenergic neurones to axotomy and nerve growth factor. Brain Res. *94:* 87–97 (1975).
9 *Levi-Montalcini, R. and Angeletti, P.U.:* Nerve growth factor. Physiol. Rev. *48:* 534–569 (1968).
10 *Levi-Montalcini, R.; Aloe, L.; Mugnaini, E.; Oesch, F., and Thoenen, H.:* Nerve growth factor induces volume increase and enhances tyrosine hydroxylase synthesis in chemically axotomized sympathetic ganglia of newborn rats. Proc. natn. Acad. Sci. USA *72:* 595–599 (1975).
11 *Lowry, O.H.; Rosebrough, N.H.; Farr, A.L., and Randall, R.J.:* Protein measurement with the Folin phenol reagent. J. biol. Chem. *193:* 265–275 (1951).
12 *Molinoff, P.B.; Weinshilboum, R., and Axelrod, J.:* A sensitive enzyme assay for dopamine β-hydroxylase. J. Pharmac. exp. Ther. *178:* 425–431 (1971).
13 *Molinoff, P.B. and Axelrod, J.:* Biochemistry of catecholamines. A. Rev. Biochem. *40:* 465–500 (1971).
14 *Mueller, R.A.; Thoenen, H., and Axelrod, J.:* Increase in tyrosine hydroxylase activity after reserpine administration. J. Pharmac. exp. Ther. *169:* 74–79 (1969).
15 *Otten, U.; Schwab. M.; Gagnon, C., and Thoenen, H.:* Selective induction of tyrosine hydroxylase and dopamine β-hydroxylase by nerve growth factor: comparison between adrenal medulla and sympathetic ganglia of adult and newborn rats. Brain Res. (in press).
16 *Ouchterlony, O.:* Handbook of immunodiffusion and immunoelectrophoresis (Ann Arbor Scientific Publishers, Ann Arbor 1970).
17 *Purves, D. and Njå, A.:* Effects of nerve growth factor on synaptic depression after axotomy. Nature, Lond. *260:* 535–536 (1976).
18 *Stöckel, K.; Gagnon, C.; Guroff, G., and Thoenen, H.:* Purification of NGF-antibodies by affinity chromatography. J. Neurochem. *26:* 1207–1211 (1976).
19 *Thoenen, H.; Angeletti, P.U.; Levi-Montalcini, R., and Kettler, R.:* Selective induction by nerve growth factor of tyrosine hydroxylase and dopamine β-hydroxylase in the rat superior cervical ganglia. Proc. natn. Acad. Sci. USA *68:* 1598–1602 (1971).
20 *Thoenen, H.:* Comparison between the effect of neuronal activity and nerve growth factor on the enzymes involved in the synthesis of norepinephrine. Pharmac. Rev. *24:* 255–267 (1972).

U. Otten, MD, Biocenter of the University Basel, *CH–4000 Basel* (Switzerland)

Some Aspects of GABA Level Regulation in Developing Rat Brain

L. Ossola, M. Maitre, J.M. Blindermann and P. Mandel

Centre de Neurochimie and Institut de Chimie Biologique, Faculté de Médecine, Strasbourg

Introduction

γ-Aminobutyric acid (GABA) is the major inhibitory neurotransmitter in the central nervous system (3, 11, 19). Previous studies in various species have indicated that levels of GABA in brain are relatively high at early stages of development (4, 17, 18, 21, 22). This is in agreement with electrophysiological studies suggesting that tonic inhibitory influences predominate in the brain over excitatory ones, even early in development (10). However, glutamic acid decarboxylase (GAD) activity in the newborn is about 10% of that in the adult and the level of GABA at birth is high even if the amount of its biosynthetic enzyme in different brain regions is rather low (8). This differs markedly from the noradrenergic system where in the whole brain there is a relatively close association between the developmental increase in the activity of tyrosine hydroxylase and dopamine β-hydroxylase and the level of endogenous norepinephrine (6, 7, 9). This discrepancy between the amount of GABA and GAD might be explained by regulation at the level of transport or GABA degradation. We have investigated this last possibility by measuring the molecular activities of the GABA catabolizing enzyme, GABA aminotransferase (GABA-T), in the developing rat brain, in parallel with GABA levels.

Materials and Methods

Preparation of Tissues

Experiments were performed on Wistar rats. After parturition, litter sizes were reduced to a maximum of 6 pups. For measurements of the enzymatic activities and for radioimmunoassays, animals were killed by decapitation and tissues were homogenized at 4 °C in 6 vol (w/v) of 20 mM phosphate buffer pH 7.4 containing 0.5% Triton X-100 (v/v), 0.5 mM pyridoxal phosphate, 1 mM 2-aminoethylisothiouronium bromide hydrobromide and 0.1 mM EDTA. For the determination of GABA, the animals were killed in a microwave oven. The irradiation time of the head of the animals never exceeded 7 sec to obtain a denaturation of enzymes directly involved in GABA metabolism. The brains were homogenized in 10 vol of ice cold 80% ethanol and centrifuged at 10,000 g for 30 min at 4 °C. The pellet was re-extracted with 10 vol of ice-cold 70% ethanol and re-centrifuged. The supernatants were pooled and evaporated to dryness under low pressure at room temperature.

Protein concentrations were determined by the method of *Lowry et al.* (15) with bovine serum albumin as a standard.

Assay for GAD

GAD activity was measured by using C_1 carboxy-labelled glutamate according to a modification of the method of *Albers and Brady* (1). 50 μl tissue homogenate (1–1.5 mg protein) were added into 6-ml conical test tubes containing 50 μl incubation medium to a final concentration of 5 mM glutamate (2 μCi/5 μmol), 1 mM 2-aminoethylisothiouronium bromide hydrobromide, 0.1 mM EDTA and 0.5 mM pyridoxal phosphate, 100 mM potassium phosphate buffer pH 6.8. Each assay was done in triplicate and incubated at 37 °C for 40 min. Blanks consisted of the standard incubation mixture alone. Enzyme activity was quantified by measuring the amount of $^{14}CO_2$ evolved and absorbed by 100 μl of Hyamine hydroxide.

Assay for GABA-T

Incubations were carried out at 37 °C in a total volume of 70 μl 0.3 M Tris-HCl buffer (pH 8.6) containing the following reagents (final concentration): 2-ketoglutaric acid, 4 mM; 2-aminoethylisothiouronium bromide hydrobromide 1 mM; EDTA, 0.1 mM; pyridoxal phosphate, 0.5 mM; sodium succinate, 0.5 mM; NAD, 2.5 mM; [^{3}H 2–3] GABA, 8 mM (8 μCi/μmol). The reaction was started by the addition of 20 μl tissue homogenate and stopped after 40 min by adding 100 μl trichloroacetic acid (10% w/v). Blanks were run under the same conditions but without tissue homogenate. The incubation medium was applied to a 0.5 × 3 cm column of Dowex 50 W × 8 resin (H$^+$ form). The succinic acid product was eluted with 4 times 1 ml of water. The eluates were collected in 20-ml scintillation vials, containing 3 ml of Instagel and counted in an Intertechnique scintillation spectrometer. Each determination was run in triplicate.

Radioimmunoassay for GABA-T

GABA-T from rat brain was purified as previously described and homogeneity studies on such preparations have been demonstrated (16). The ^{125}I-labelled GABA-T was prepared weekly by the method of *Hunter* (13) using chloramine-T, except that buffering capacity was increased by the addition of 100 μl of 0.5 M phosphate buffer pH 7.5. Specific radioactivities ranged from 6 to 8 μCi/mg.

Antiserum to rat GABA-T was obtained in an adult rabbit 45 days after three injections of 95% pure GABA-T (each of 0.5 mg protein). Freund's complete adjuvant was used as a vehicle for the three subcutaneous injections separated by an interval of 15 days. The antiserum was stored at 4 °C with 1% sodium azide as preservative.

Immunodiffusion and immunoelectrophoresis were carried out with rabbit anti GABA-T serum against a 125-fold purified enzyme obtained by ammonium sulfate fractionation (16). One precipitin band was obtained by immunodiffusion and immunoelectrophoresis.

Titration of Antiserum

Dilutions of antiserum were made in 20 ml phosphate buffer (pH 7.4) containing 1% bovine serum albumin, 0.1% sodium azide and 0.5% Triton X-100. The ^{125}I-GABA-T (0.2 ml of a solution containing about 0.02 µg/ml) was incubated with 0.2 ml antiserum at a final dilution ranging from 1/500 to 1/20,000. Serum from nonimmunized rabbits was added to all dilutions such that the final concentration of serum (nonimmunized + immunized) was 1/500 in all media. Incubation was carried out at 37 °C and 6 h later, 0.2 ml of a 1/300 dilution of goat anti-rabbit γ-globulin was added in all samples. A new incubation was carried out for 15 h at 4 °C. The precipitate formed, containing the radioactivity of the ^{125}I-labelled GABA-T bound to the immune γ-globulin anti-GABA-T was isolated by filtration on GS Millipore filter 0.22 µm.

Inhibition Test

This test was done with the dilution of antibodies that give 60–80% binding of antigen. Unlabelled GABA-T was incubated in a range from 0 to 0.04 µg/400 µl in the presence of a constant dilution of antibody and a constant level of labelled GABA-T. The incubation medium consisted of phosphate buffer 20 mM (pH 7.4) containing (final concentration): 1% bovine serum albumin, 0.1% sodium azide, 0.5% Triton X-100 and 1/500 rabbit serum (immunized + normal). The same procedure as for the titration of antiserum was utilized and in parallel experiments, brain homogenate (5–20 µg protein) was substituted to unlabelled GABA-T.

Assay of GABA

The GABA content in the extract, redissolved with a minimum volume of 0.01 N HCl, was determined by the enzymatic fluorometric assay of *Graham and Aprison* (12). Since some constituents of the extract interfered with the fluorometric assay for GABA, the method of standard addition was employed (20). The assays were run in the same conditions as the calibration curves.

Results

Glutamate Decarboxylase in Developing Rat Brain

During the postnatal period, the activity of glutamate decarboxylase increases about 4–5 times in brain hemispheres while in the cerebellum, the activity remains unchanged until 10 days of age and then increases slightly (fig. 1).

GABA Concentration in Developing Rat Brain

The level of GABA per milligram protein or per milligram wet weight in brain hemispheres remains almost constant from birth to adult age (fig. 2). In cerebellum there is a slight decrease in GABA level per milligram protein from the first week postpartum to adult age, while the value per milligram wet weight remains unchanged (fig. 2).

GABA-T Activity as a Function of Protein Dilution

The specific activity of GABA-T decreases with the increase of the protein concentration in the incubation medium (5, 23). The observed GABA-T activity in the homogenate increases linearly until a concentration of 2.8 mg protein/ml

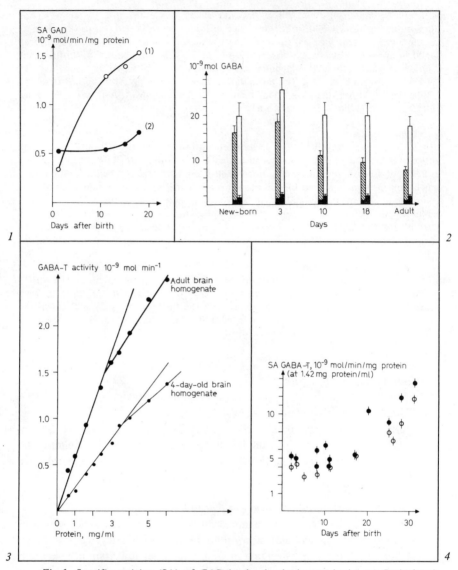

Fig. 1. Specific activity (SA) of GAD in the developing rat brain. ○ = Brain hemispheres, ● = cerebellum.

Fig. 2. Brain level of GABA in the developing rat brain. ■ = Per milligram wet weight, □ = brain hemispheres, ▨ = cerebellum.

Fig. 3. GABA-T activity as a function of the protein concentration.

Fig. 4. Specific activity (SA) of GABA-T in the developing rat brain. Measurements were made with a constant dilution of protein at pH 8.6 and at 37 °C. ○ = Brain hemispheres; ● = cerebellum.

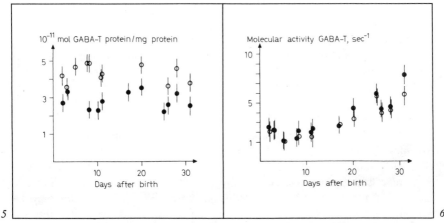

Fig. 5. GABA-T molecules per milligram protein in the developing rat brain. Measurements were made by the radioimmuno technique. ● = Brain hemispheres; ○ = cerebellum.

Fig. 6. Molecular activity of GABA-T in the developing rat brain. Results expressed as number of GABA molecules transformed at 37 °C and pH 8.6 in 1 sec per molecule of GABA-T. ● = Brain hemispheres; ○ = cerebellum.

for the adult and 3.8 mg protein/ml for the newborn is reached. Then a break in the slope occurs and the observed activity of GABA-T decreases (fig. 3). This indicates that the protein concentration must be kept constant when comparing the specific activity of GABA-T in different brain homogenates.

GABA Transaminase Activity in Developing Rat Brain

In the cerebellum as well as in brain hemispheres, an increase of the GABA-T specific activity occurs during development (fig. 4). This specific activity was always measured with a constant protein concentration of 1.42 mg/ml and increased about 3 times 30 days after birth.

Level of GABA Transaminase during Development of the Rat Brain

Figure 5 shows that the level of the enzyme per milligram protein remains almost constant from birth to 30 days old. The same results were obtained for brain hemispheres and for the cerebellum.

Molecular Activity of GABA-T in Developing Rat Brain

Molecular activity of GABA-T was expressed as the number of molecules of GABA transformed per second at 37 °C and pH 8.6 per molecule of GABA-T. As shown in figure 6, this number increases in both brain hemispheres and in cerebellum from 2 to 8 and 2 to 6 sec^{-1}, respectively.

Discussion

In contrast to a previous report (8), we found that GABA per milligram protein remains almost constant in brain hemispheres from birth to adult age. A slight increase is observed at 5 days of age. This discordance with other results is probably due to the different method of sacrifice. In previous studies, the animals were sacrificed by decapitation followed by freezing. It is possible that the postmortem increase of the GABA level was proportional to the tissue volume to be freezed, which increased with age. The higher activity of the enzymes GAD and GABA-T in adult brain favors the postmortem increase in GABA (2).

There is a marked disparity between the increase of GAD and the maintenance of GABA at an almost constant level. This phenomenon may be explained by a change in activity of GABA-T and/or GABA transport through the cell membranes. Also the high affinity uptake of GABA by the whole brain seems to remain unchanged during development (8). However, one must keep in mind that GABA is also an intermediate in the metabolism of glucose and that there is a pool of GABA which is not directly involved in neurotransmission.

GABA-T activity increased during brain development in the rat. It has been reported that in partially purified brain GABA-T (23) or in brain homogenate (5), there is a decrease in specific activity as a function of enzyme or tissue concentration. This phenomenon has been explained by a tendency of the enzyme to aggregate. Our results suggested that the increase in GABA-T activity in rat brain during development is due to the modulation of the level of a GABA-T effector. Molecular activity of GABA-T had been measured by spectrophotometric method (14) and a value of 9.5 sec^{-1} was obtained. The molecular activity of GABA-T in homogenate, measured with our technique, is always below this value, suggesting that a GABA-T inhibitor is present in the homogenate. The presence of such an inhibitor might explain the change of GABA-T activity observed at high tissue concentration. It explains also the absence of change of GABA-T molecules per milligram protein measured by radioimmunoassay during development while GABA-T activity increases. This phenomenon might be ascribed to a decrease of the inhibitor at a later age.

Summary

The specific activities of the GABA synthetizing and degrading enzymes, glutamate decarboxylase and GABA transaminase, increase during postnatal development in rat brain but the level of GABA remains constant in brain hemispheres and in the cerebellum. The specific activity of GABA transaminase increases with the dilution of the homogenate, thus, the enzymatic activities were determined at a constant dilution. A radioimmunoassay, allowing the measurement of the protein GABA-T, gives more precise information. Actually, the quantity of enzyme per milligram of protein remains constant during the development of the brain. Therefore the observed increase in specific activities of GABA-T in brain seems to indicate the presence of an effector of the GABA-T activity and it appears that modulation of the level of this effector but not of the level of GABA-T protein modifies the total GABA-T activity during growth. This mechanism contributes to the regulation of the GABA level in the developing rat brain.

References

1 *Albers, R.W. and Brady, R.O.:* The distribution of glutamate decarboxylase in the nervous system of the rhesus monkey. J. biol. Chem. *234:* 926–928 (1959).
2 *Balcom, G.J.; Lenox, R.H., and Meyerhoff, J.L.:* Regional gamma-aminobutyric acid levels in rat brain determined after microwave fixation. J. Neurochem. *24:* 609–613 (1975).
3 *Baxter, C.F.:* The nature of gamma-aminobutyric acid. Handbook of neurochemistry, vol. 3, pp. 289–353 (Plenum Press, New York 1969).
4 *Baxter, C.F. and Roberts, E.:* Gamma-aminobutyric acid and cerebral metabolism; in *Brady and Tower* The neurochemistry of nucleotides and amino acids, pp. 127–145 (Wiley, New York 1960).
5 *Boer, T. de and Bruinvels, J.:* Assay and properties of 4-aminobutyric 2-oxoglutaric acid transaminase and succinic semialdehyde dehydrogenase in rat brain tissue. J. Neurochem. *28:* 471–478 (1977).
6 *Coyle, J.T.:* Biochemical aspects of catecholaminergic neurons in the brain of the fetal and neonatal rat; in *Fuxe, Olsen and Zotterman* Dynamics of degeneration and growth in neurons, pp. 425–434 (Pergamon Press, New York 1974).
7 *Coyle, J.T. and Axelrod, J.:* Tyrosine hydroxylase in rat brain: developmental characteristics. J. Neurochem. *19:* 1117–1123 (1972).
8 *Coyle, J.T. and Enna, S.J.:* Neurochemical aspects of the ontogenesis of gabanergic neurons in the rat brain. Brain Res. *111:* 119–133 (1976).
9 *Coyle, J.T. and Henry, D.:* Catecholamines in fetal and newborn rat brain. J. Neurochem *21:* 61–67 (1973).
10 *Crain, S.M. and Bornstein, M.B.:* Early onset in inhibitory functions during synaptogenesis in fetal mouse brain cultures. Brain Res. *68:* 351–357 (1974).
11 *Curtis, D.R. and Johnston, G.A.R.:* Amino acid transmitters in the mammalian central nervous system. Ergebn. Physiol. *69:* 97–188 (1974).
12 *Graham, L.T. and Aprison, M.H.:* Fluorometric determinations of aspartate, glutamate and gamma-aminobutyric acid in nerve tissue using enzymatic methods. Analyt. Biochem. *15:* 487–497 (1966).
13 *Hunter, W.M.:* Preparation and assessment of radioactive tracers. Br. med. Bull. *30:* 18–23 (1974).
14 *John, R.A. and Fowler, L.J.:* Kinetic and spectral properties of rabbit 4-aminobutyrate amino-transferase. Biochem. J. *155:* 645–651 (1976).
15 *Lowry, O.H.; Rosebrough, N.J.; Farr, A.L., and Randall, R.J.:* Protein measurement with the Folin phenol reagent. J. biol. Chem. *193:* 265–275 (1951).
16 *Maitre, M.; Ciesielski, L.; Cash, C., and Mandel, P.:* Purification and studies on some properties of the 4-aminobutyrate-2-oxoglutarate transaminase from rat brain. Eur. J. Biochem. *52:* 157–169 (1975).
17 *Roberts, E.:* Metabolism of gamma-aminobutyric acid in various areas of brain; in *Kety and Elkes* Regional neurochemistry, pp. 324–347 (Pergamon Press, New York 1961).

18 *Roberts, E.; Harman, P.J., and Frankel, S.:* Gamma-aminobutyric acid content and glutamic decarboxylase activity in developing mouse brain. Proc. Soc. exp. Biol. Med. *78:* 799–803 (1951).
19 *Storm-Mathisen, J.:* Gaba as a transmitter in the central nervous system of vertebrates. J. neural Trans., suppl. XI, pp. 227–253 (1974).
20 *Vallee, B.L. and Thiers, R.E.:* Flame photometry. Treat. analyt. Chem. *6:* 3463–3475 (1965).
21 *Van den Berg, C.J.; Van Kempen, G.M.J.; Schade, J.P., and Veldstra, M.:* Levels and intracellular localization of glutamate decarboxylase and other enzymes during the development of the brain. J. Neurochem. *12:* 863–869 (1965).
22 *Vernadakis, A. and Woodbury, D.M.:* Electrolyte and amino acid changes in rat brain during maturation. Am. J. Physiol. *203:* 748–752 (1962).
23 *Waksman, A. and Roberts, E.:* Purification and some properties of mouse brain gamma-aminobutyric-α-ketoglutaric acid transaminase. Biochemistry *4:* 2132–2138 (1965).

Dr. *L. Ossola,* Centre de Neurochimie, Faculté de Médecine, 11, rue Humann, *F–67085 Strasbourg Cedex* (France)

One Molecular Form of AChE Associated with Synapses in Two Cholinergic Systems: Skeletal Muscle of the Rat and Ciliary Ganglion of the Chick[1]

Jeanine Koenig and Herbert L. Koenig

Laboratoire de Neurocytologie, Université P. et M. Curie, and Inserm, U-153, Paris

Introduction

The majority of the studies concerning acetylcholinesterase (AChE) were performed by physiological and histochemical methods. Electrophysiological methods were used to establish the physiological role of AChE in the transmission of the nerve impulse at cholinergic synapses. The enzyme hydrolyses the neurotransmitter ACh released by presynaptic nerve endings, terminates its action on the postsynaptic membrane and prevents desensitization of the ACh receptors. Cytochemistry permitted localization of AChE activity at cellular level. In muscle fibers, most of the enzyme is concentrated at the motor end-plates, although it is also detected in areas of the fibers lacking end-plates and at myotendinous junctions (4). In the autonomic nervous system, it is a generally accepted view that in certain ganglia, two sites of AChE activity exist. One is localized in presynaptic structures, as shown by its disappearance after section of the presynaptic axons and the other in the postsynaptic cell bodies (17).

Recently, a progress in the knowledge of the biochemical properties of AChE was introduced by *Massoulie and Rieger* (11). They observed that in crude extracts of electric organs of *Electrophorus electricus* and *Torpedo marmorata,* it is possible to distinguish several molecular forms of AChE, by their sedimentation constants in a sucrose gradient. The existence of molecular forms of AChE has since been described in various mammalian (6a, 12) and avian tissues (15).

---

[1] This work was done in collaboration with: *M. Vigny* for the muscle; *J.-Y. Couraud* and *L. Di Giamberardino* for the chicken ciliary ganglion, and *M. Vigny* and *J. Massoulie* for ciliary ganglia cultures.

This work was supported by funds from the Institut National de la Santé et de la Recherche Médicale 5266 ATP 3876-70 and 77.1.180.6.B.

The question which immediately arose was to correlate these forms with particular structures and/or functions. The first evidence of a possible correlation between one molecular form and a specific structure was obtained in the diaphragm of the rat. *Hall* (6a) established that the heaviest form of AChE (16 S) is specifically associated with end-plate regions of the muscle, and suggested that it could correspond to the end-plate enzyme. This hypothesis was substantiated by an analysis of the behavior of the 16 S form in rat muscles taken in various experimental conditions (14).

The results of these experiments demonstrate that the 16 S form of AChE is actually detected in every situation where the muscle fibers contain neuromuscular junctions: in the end-plate zones, when the old end-plates are reinnervated after denervation and when new end-plates are formed at ectopic sites. In contrast, 16 S is not present in muscles deprived of end-plates: nerve-free segments and late denervated muscles.

A heavy molecular form of AChE is also detected in two autonomic ganglia, the cervical superior ganglion of the rat (6b) and the ciliary ganglion of the chick (15).

The aim of this investigation is (1) to determine if the heavy form of AChE is associated with synapses in the ciliary ganglion of the chick, a purely cholinergic ganglion (20 S form), and (2) to study the correlation between the presence of the heavy forms of AChE and synaptogenesis in the two cholinergic systems, the muscle cells and the ganglionic cells.

Muscle of the Rat

During embryogenesis, the 16 S form of AChE is not detected in hindleg muscles of the rat embryos before 14–15 days of gestation (14). Unfortunately, there is neither morphological nor electrophysiological evidence that synapses are formed at this stage of development in these muscles.

In order to obtain detectable synapses with certainty, we cultivated myoblasts and coupled them with spinal cord neurones (10). As the myoblasts were obtained from muscles of 13- to 14-day embryos which are devoid of 16 S AChE, it was possible to establish that the appearance of this form in the cultures was induced by the presence of the neurones (fig. 1). The 16 S form appears 6 days after the neurones are added to the myoblasts, and is maintained in the cultures for 3–4 weeks, until the cells peeled from the dishes. Myoblasts cultivated without neurones and spinal cord cultures never show a 16 S activity, even in long-lasting cultures. To determine if the neural induction of the 16 S form is mediated through axon terminals and if synapses are formed at this time, we used the Koelle histochemical method for AChE as a marker of synaptic sites (7). It is generally accepted that localized concentrations of AChE activity occur

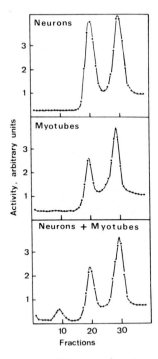

Fig. 1. AChE molecular forms in cell cultures plated for 6 days. In neurones and myotube cultures, the two peaks are: left = 10 S, right = 4 S. In neuron + myotube cultures, in addition to 4 S and 10 S, note the small peak of 16 S AChE.

only when the contact between axons and muscle cells is established (1, 9). In cultures coupled for 6 days, the AChE activity is concentrated at the periphery of spot-like areas (fig. 2A, B). This localization of the deposits suggests that the AChE spots correspond to shallow depressions of the sarcolemma induced by varicosities of the axons applied on the myotubes (fig. 2D). In 10-day coupled cultures, the spots are replaced by localizations typical of end-plates (fig. 2C) and similar to the newly formed end-plates obtained when a motor nerve is implanted in a nerve-free segment of muscle (9). It is likely that in our 10-day coupled cultures, as in newly formed end-plates, the neuromuscular junctions are immature and in particular lack subneural infoldings (9). The rapid change in the shape of AChE deposits indicates that the AChE spots observed at 6 days represent the earliest neuromuscular contacts.

Our results show that the presence of neurons in the cultures is not sufficient to induce the 16 S form of AChE and that close contacts between presynaptic axons and muscle cells are necessary.

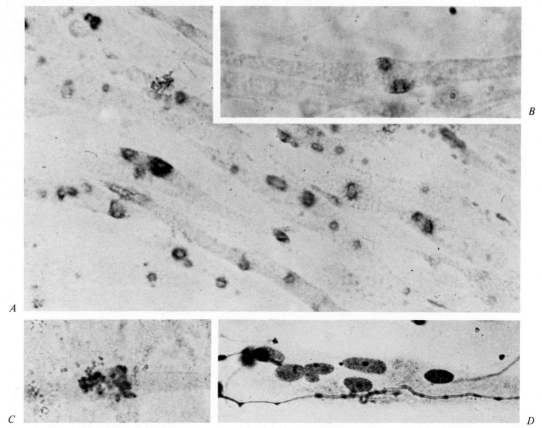

Fig. 2. Coupled cultures of myoblasts and spinal cord neurons. *A, B* 5 days coupling. Observe the spots of high AChE activity on the membranes of young myotubes. Histochemical method of *Koelle.* × 800. *C* 10 days coupling. End-plates formed by short gutters. Histochemical method of *Koelle.* × 800. *D* 5 days coupling. Axon varicosities applied on the muscular membrane. Silver impregnation. × 800.

Ciliary Ganglion of the Chick

In this ganglion, four molecular forms of AChE were identified (15). According to the results obtained in muscle, it seemed likely that, if a peculiar form were associated with synapses, it would be the heaviest. Thus, we determined the time of appearance of the 20 S form in embryonic ganglia.

In ganglia of 7-day-old embryos (stage 31–32), two molecular forms are distinguished: 7.5 S and 11.5 S, which represent 90 and 10% of the total AChE activity. Although numerous ganglionic cells are contacted by presynaptic axon

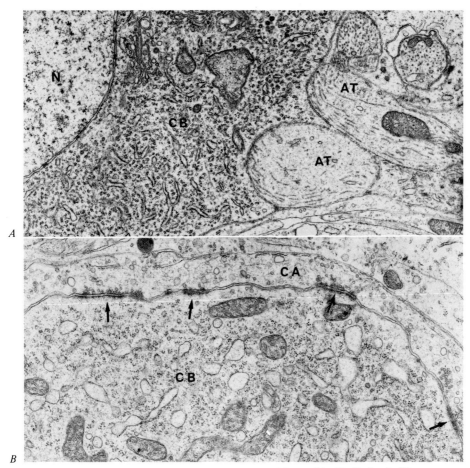

Fig. 3. Ciliary ganglion of chicken embryos. *A* Stage 31–32 (7 days). The axon terminals (AT) contain microtubules, filaments, mitochondria and SER. No membrane specializations nor synaptic vesicles are observed. CB = Cell body; N = nucleus. × 23,000. *B* Stage 36 (10 days). The presynaptic nerve endings (CA) (calyx) encompass the cell body (CB). Note the synapses (arrows). × 29,000.

terminals, 20 S activity is not detected. Axosomatic synapses with membrane specializations and synaptic vesicles in the presynaptic terminals were not observed at this stage (fig. 3A). In ganglia of 10-day-old embryos (stage 36), characteristic presynaptic nerve endings and synapses are formed on the postsynaptic cell bodies (fig. 3B). In these ganglia, the amount of 20 S is the same as in young chicken ganglia, 10%.

In cell cultures of ciliary ganglion coupled with spinal cord, we were unable to detect synapses or 20 S AChE. Two sites of AChE activity could be detected histochemically in the ciliary ganglion of the chick: the presynaptic axons and nerve endings and the postsynaptic cell bodies and their axons (8, 16). To determine if the 20 S form is associated either with the pre- or the postsynaptic structures, we analyzed the behavior of this molecular form in three experimental conditions (fig. 4). In denervated ganglia, obtained by section of the preganglionic nerve, the total AChE activity decreases by 30% in 2 days (5, and unpublished). During the same period, the 20 S form declines by 70%. From this data one would expect a 30% postsynaptic localization of this form. In fact, when the postganglionic nerves are cut (axotomy), the 20 S activity is reduced by 60% in 3 days but the total AChE activity remains 80% of the control (5, and unpublished).

The massive decrease of the 20 S form, after both pre- and postganglionic nerve sections, excludes to attribute a definite site, either pre- or postsynaptic, to this form. Its maintenance appears to be dependent of the structural and functional integrity of the synapses. The almost complete disappearance of 20 S after double sections, when pre- plus postganglionic nerves are sectioned simultaneously (fig. 4C), shows that a reciprocal modulation of this particular form occurs, as *Chiappinelli et al.* (3) postulated for the total ganglionic AChE activity. Furthermore, preganglionic sections induce a parallel and significant decrease of 20 S in the controlateral ganglion (fig. 4A, C). One possible explanation for this effect would be that the degeneration of the nerve endings after the preganglionic nerve section affects the functional state of the contralateral presynaptic nerve endings, through an unknown mechanism, and thereby on the level of the 20 S AChE associated with them.

Discussion

In the two cholinergic systems studied it appears that the heaviest molecular form of AChE, 16 S in rat muscle and 20 S in chicken ciliary ganglion, is closely associated with synapses. Thus, it may be referred to as the 'synaptic' form of this enzyme. The appearance of the heavy form is induced in muscle and in ganglionic cells, under neural influence and it coincides with morphologically well-established contacts between presynaptic axons and target cells. More

---

Fig. 4. Kinetics of the 20 S form of AChE in ciliary ganglion of the chick in three experimental conditions. *A* Denervation (section of preganglionic nerve). *B* Axotomy (section of postganglionic nerves). *C* Double section (pre- plus postganglionic nerves are sectioned). Activity is expressed as nmoles of acetylthiocholine per minute per ganglion; 3–5 ganglia were measured for each point of the curves. ● = Operated side ganglia; ■ = controlateral ganglia; ▲ = ganglia of non-operated chicken.

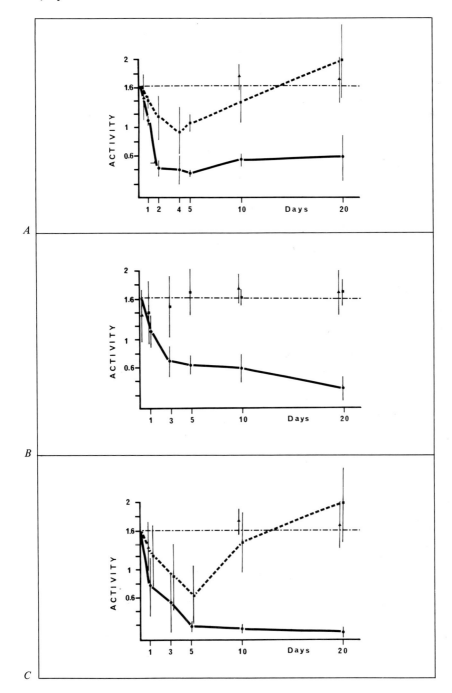

precisely, characteristic synapses with membrane specializations are formed at this time on the ganglionic cell bodies. It is likely that in the earliest neuromuscular contacts some specializations are also present. Once it is induced, the maintenance of the so-called 'synaptic' form of AChE remains under neural control and depends, in ciliary ganglion at least, on the functional state of the synapses. It seems possible to postulate that the heavy form is the unique form of the enzyme AChE directly involved in transmission. According to this hypothesis, only a small part of the enzyme (10% in ciliary ganglion and 5% in muscle) would be necessary for the degradation of ACh.

Several questions of interest remain unsolved: (1) is the signal which induces the 'synaptic associated' form the same as for the other forms?, (2) is this form constructed by the assembly of the lighter forms of AChE as suggested by recovery experiments after DFP treatment (10, 13, 17)?, (3) is the 20 S form in the ciliary ganglion synthesized in the postsynaptic cell body, as in the muscle cell, or is it conveyed to the synapse with the axoplasmic transport in presynaptic axons?, and (4) what is the precise localization of the 'synaptic' form? Is it, as postulated for the whole AChE, associated with the synaptic cleft material (2)?

Summary

A previous experimental analysis in the rat muscle has shown that one molecular form of AChE (16 S) is associated with the presence of neuromuscular junctions. Evidence is presented here that in myogenic cell cultures, this particular form is induced under neuronal influence. At the time 16 S appears, synaptic sites are visualized on the surface of the myotubes, as revealed by the presence of AChE deposits and axonal varicosities. In the ciliary ganglion of chick embryos, a parallel EM and biochemical analysis shows that, during synaptogenesis, a heavy form of AChE (20 S) appears only when the presynaptic nerve endings with characteristic synapses are detected on the ganglion cells. The maintenance of this AChE form in the chick is shown to be dependent on the structural and functional integrity of the synapses. It is concluded that in both cholinergic systems, the heaviest molecular form of AChE may be referred to as the 'synaptic form' of the enzyme. It is postulated that this could be the unique form of the enzyme directly involved in transmission.

References

1. Bennet, M.R. and Pettigrew, A.G.: The formation of neuromuscular synapses. Cold Spring Harb. Symp. quant. Biol. *40:* 409–424 (1976).
2. Betz, W. and Sakman, B.: Effects of proteolytic enzymes on function and structure of frog neuromuscular junctions. J. Physiol., Lond. *230:* 675–688 (1973).
3. Chiappinelli, V.; Giacobini, E.; Pilar, G., and Uchimura, H.: Induction of cholinergic enzymes in chick ciliary ganglion and iris muscle cells during synapse formation. J. Physiol., Lond. *257:* 749–766 (1976).

4   *Couteaux, R.:* Structure and cytochemical characteristics of the neuromuscular junction. International encyclopaedia of pharmacology and therapeutics, section 14, vol. 1, pp. 7–56 (Pergamon Press, New York 1972).
5   *Giacobini, E. and Chiappinelli, V.:* The ciliary ganglion: a model of cholinergic synaptogenesis; in *Tauc* Synaptogenesis. Gif-lectures in neurobiology (Naturalia and Biologica, 1976).
6a *Hall, Z.:* Multiple forms of acetylcholinesterase and their distribution in end plate and non endplate regions of rat diaphragm muscle. J. Neurobiol. *4:* 343–361 (1973).
6b *Gisiger, V.; Vigny, M.; Gautron, J., and Rieger, F.:* Acetylcholinesterase of rat sympathetic ganglion. Molecular forms, localization and effect of denervation. J. Neurochem. (in press).
7   *Koelle, G.B. and Friedenwald, J.S.:* A histochemical method for localizing cholinesterase activity. Proc. Soc. exp. Biol. Med. *70:* 617–622 (1949).
8   *Koenig, H.L.:* Relations entre la distribution de l'activité acétylcholinestérasique et celle de l'ergastoplasme dans les neurones du ganglion ciliaire du poulet. Archs Anat. microsc. Morph. exp. *54:* 937–964 (1965).
9   *Koenig, J. et Pecot-Dechavassine, M.:* Relations entre l'apparition des potentiels miniatures spontanés et l'ultrastructure des plaques motrices en voie de réinnervation et de néoformation chez le rat. Brain Res. *27:* 43–57 (1971).
10 *Koenig, J. and Vigny, M.:* Neural induction of the 16 S molecular form of AChE in muscle cell cultures. Nature, Lond. *271:* 75–77 (1978).
11 *Massoulie, J. et Rieger, F.:* L'acétylcholinestérase des organes électriques de poissons (Torpille et Gymnote): complexes membranaires. Eur. J. Biochem. *11:* 441–455 (1969).
12 *Rieger, F. and Vigny, M.:* Solubilization and physicochemical characteristics of rat brain AChE: development and maturation of its molecular forms. J. Neurochem. *25:* 121–129 (1976).
13 *Rieger, F.; Bauman, N.; Benda, P., and Vigny, M.:* Molecular forms of acetylcholinesterase; their *de novo* synthesis in mouse neuroblastoma cells. J. Neurochem. *27:* 1059–1063 (1976).
14 *Vigny, M.; Koenig, J., and Rieger, F.:* The motor endplate specific form of acetylcholinesterase: appearance during embryogenesis and reinnervation of rat muscle. J. Neurochem. *27:* 1347–1353 (1976).
15 *Vigny, M.; Di Giamberardino, L.; Couraud, Y.; Rieger, F., and Koenig, J.:* Molecular forms of chicken acetylcholinesterase: effect of denervation. FEBS Lett. *69:* 279–280 (1976).
16 *Taxi, J.:* Contribution à l'étude des connexions des neurones moteurs du système nerveux autonome. Annls Sci. nat., Zool. *7:* 413–674 (1965).
17 *Wilson, W. and Walker, C.R.:* Regulation of newly synthetized AChE in muscle cultures treated with DFP. Ann. N.Y. Acad. Sci. *71:* 3194–3198 (1974).

Dr. *J. Koenig,* Université P. et M. Curie, Laboratoire de Neurocytologie, 12, rue Cuvier, *F–75005 Paris* (France)

Multiple Forms of Monoamine Oxidase in the Human Cerebral Cortices at Different Ages

O. Suzuki and K. Yagi

Institute of Biochemistry, Faculty of Medicine, University of Nagoya, Nagoya

Introduction

Recently, many studies have been centered around multiple forms of monoamine oxidase [amine: oxygen oxidoreductase (deaminating) (flavin-containing); EC 1.4.3.4](MAO). In these studies, the existence of two types of MAO, which are designated as type A and type B enzyme, has been well documented (5, 19, 20). Even though it is not clear if each type of MAO has its own enzyme protein, it is unequivocally accepted that both types of MAO are demonstrable both *in vitro* and *in vivo* in their substrate specificity and inhibitor sensitivity. Type A MAO is sensitive to the inhibitor drug, clorgyline, and 5-hydroxytryptamine (5-HT) as well as norepinephrine is preferred as substrate for this enzyme. Type B MAO is sensitive to the inhibitor drug, deprenyl, and β-phenylethylamine (PEA) as well as benzylamine is preferred as substrate. Some substrates such as tyramine, tryptamine, dopamine and kynuramine are oxidized by either type of MAO.

In the physiological aspect of the studies on MAO, the developments of the two types of MAO in the brain have been the focus of interest (2, 6, 10, 21). However, the studies have been limited to rats and mice. The present paper deals with the two types of MAO in the human cerebral cortices at different ages.

Materials and Methods

Materials

[2-^{14}C]5-HT creatinine sulphate (58 mCi/mmol) was purchased from the Radiochemical Centre, Amersham. 5-HT creatinine sulphate, PEA-HCl, benzylamine-HCl, chloral hydrate and Triton X-100 were obtained from Nakarai Chemicals, Kyoto; 4-hydroxyquinoline from Tokyo Kasei Kogyo Co., Tokyo; kynuramine-2HBr from Sigma Chemical

Co., St. Louis, Mo.; phenylacetaldehyde from Aldrich Chemical Company, Milwaukee, Wisc.; and Sephadex G-25 from Pharmacia Fine Chemicals, Uppsala. Deprenyl was kindly donated by Prof. *J. Knoll,* Department of Pharmacology, Semmelweis University of Medicine, Budapest.

Human fetal brains were kindly donated by Dr. *T. Maruyama,* Department of Obstetrics and Gynaecology, Ekisaikai Hospital, Nagoya and Dr. *K. Matsuura,* the Matsuura Clinic of Obstetrics and Gynaecology, Nagoya. Fetal brains were secured only when therapeutic abortions were performed. Infant (3 years) and adult brains were obtained at autopsy 5–14 h after death from patients of brain cancer. The normal cerebral cortex was excised from the frontal lobe, frozen on dry ice and kept in a freezer until used. In order to check the post-mortem change in MAO, MAO activity towards kynuramine and its inhibition by 10^{-7} M deprenyl were examined with rat brain after rats were killed by decapitation and left at room temperature for 24 h. It was confirmed that neither MAO activity nor its inhibition by deprenyl in the brain was changed by its exposure to room temperature for 24 h.

Assay Procedures

MAO activity towards 5-HT was assayed by a radiochemical procedure (11). The concentration of the substrate was 250 μM. The determination of MAO activity with kynuramine as substrate was carried out fluorometrically by the method of *Kraml* (7) with a slight modification (1). The concentration of kynuramine in the assay mixture was 82 μM. MAO activity towards PEA was measured fluorometrically by our new method (15), and the concentration of the substrate was 250 μM.

The inhibition of MAO by deprenyl was studied with kynuramine as substrate according to *Squires* (13). In this experiment, the assay mixture was preincubated at 37 °C for 10 min to ensure the maximal enzyme inhibition.

All the above assays were carried out within a week after sampling. Protein in brain homogenate was determined by the biuret method (8) with a pretreatment (16), and that in the solubilized preparation by the method of *Lowry et al.* (9).

Radioactivity was measured with an Aloka LSC-502 liquid scintillation counter. All fluorometric measurements were carried out in a Shimadzu corrected spectrofluorophotometer RF-502, at 20 °C.

Solubilization of MAO

A whole fetal brain was used in this experiment. The brain was homogenized with 9 vol of 0.25 M sucrose in a Potter-Elvehjem homogenizer fitted with a Teflon pestle allowed to cool in an ice bath, and centrifuged at 1,500 g for 5 min to remove cellular debris. The resulting supernatant was centrifuged at 18,000 g for 20 min and the crude mitochondrial pellet was suspended in the sucrose solution. The suspension was re-centrifuged at 18,000 g for 30 min and the pellet was re-suspended in 10 mM sodium phosphate buffer (pH 7.2). Benzylamine was added to make a final concentration of 3 mM. The mixture was then exposed to sonication for 90 min by using a Sonifier B-10 (Branson Sonic Power Company, Danbury, Conn.) at 60 W. It was cooled with ice during sonication. After this treatment, Triton X-100 was added to make a final concentration of 2% (w/v). The sonication was continued for a further 10 min and the mixture was centrifuged at 160,000 g for 90 min at 4 °C. The clear supernatant was applied to a 1.8 × 20 cm column of Sephadex G-25 that had been equilibrated with 10 mM sodium phosphate buffer (pH 7.2), and eluted at room temperature with the same buffer solution. The eluate was subjected to the experiment for the inhibition of MAO by deprenyl.

Table I. MAO activities in the human cerebral cortices at different ages[1]

Subject	Age	Sex	MAO activity					
			5-HT		kynuramine		PEA	
			µmol/g wet weight/ 30 min	nmol/mg protein/ 30 min	µmol/g wet weight/ 30 min	nmol/mg protein/ 30 min	µmol/g wet weight/ 30 min	nmol/mg protein/ 30 min
Fetus 1	18 weeks of gestation	–	0.17	3.2	0.17	3.2	0.16	3.0
Fetus 2	18 weeks of gestation	–	0.70	13.0	0.34	6.3	0.42	7.9
Fetus 3	20 weeks of gestation	–	0.34	8.4	0.20	3.9	0.10	2.6
Fetus 4	20 weeks of gestation	–	0.17	3.3	0.10	1.7	0.09	1.8
T.H.	3 years	M	1.79	20.5	3.05	37.6	1.05	12.0
F.H.	24 years	F	1.58	14.5	3.82	35.1	1.56	14.3
M.S.	29 years	F	1.75	22.6	4.53	54.6	1.17	15.1
T.M.	54 years	F	3.04	29.3	5.14	55.6	1.72	18.8
T.K.	76 years	F	1.77	20.3	5.56	56.6	1.88	21.6

[1] The concentrations of the substrates 5-HT, kynuramine and PEA were 250, 82 and 250 µM, respectively.

Results

MAO Activities towards 5-HT, Kynuramine and PEA

MAO activities with 5-HT, kynuramine and PEA as substrate in the homogenates of the human cerebral cortices at different ages are presented in table I. When the mean values of MAO activities both per gram wet weight and per milligram protein in the four fetal cerebral cortices were compared with those in the four adult cerebral cortices, the activity of the adult was much higher than that of the fetus, which was the most remarkable with kynuramine as substrate, followed by PEA and 5-HT.

Inhibition of MAO by Deprenyl

The inhibition of MAO by deprenyl, a specific inhibitor for type B MAO, was studied with kynuramine as substrate on the homogenates of the cerebral cortices at different ages. The result is illustrated in figure 1. As can be seen from the figure, in the fetal cortex the plateau was clearly observed around 10^{-7} M of

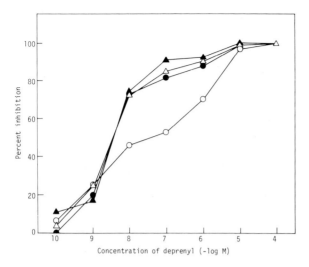

Fig. 1. Inhibition of MAO in the human cerebral cortices at different ages by deprenyl with kynuramine as substrate. ○ = Fetus at 20 weeks of gestation; ● = male child at 3 years of age (T.H.); △ = female adult at 29 years of age (M.S.); ▲ = female adult at 76 years of age (T.K.). Each point represents the mean obtained from duplicate determinations. The concentration of kynuramine was 82 μM. Homogenate of the cerebral cortex was preincubated with deprenyl at 37 °C for 10 min.

Table II. Inhibition of MAO in the human cerebral cortices at different ages by deprenyl ($10^{-7} M$)[1]

Subject	Age	Sex	Percent inhibition
Fetus 1	18 weeks of gestation	–	47.7
Fetus 2	18 weeks of gestation	–	44.5
Fetus 3	20 weeks of gestation	–	52.6
Fetus 4	20 weeks of gestation	–	58.8
T.H.	3 years	M	81.7
F.H.	24 years	F	81.3
M.S.	29 years	F	85.0
T.M.	54 years	F	85.8
T.K.	76 years	F	90.9

[1] The concentration of the substrate kynuramine was 82 μM. The homogenate of the cortex was preincubated with $10^{-7} M$ deprenyl at 37 °C for 10 min. Each value represents the mean obtained from duplicate determinations.

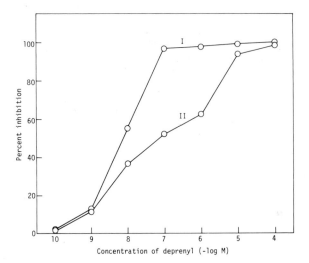

Fig. 2. Effect of solubilization of MAO in the human fetal brain on its inhibition by deprenyl. I = Solubilized enzyme preparation; II = crude mitochondrial preparation. Each point represents the mean obtained from duplicate determinations. The concentration of the substrate kynuramine was 82 μM. The enzyme preparation was preincubated with deprenyl at 37 °C for 10 min.

deprenyl, showing the biphasic responses to deprenyl. Therefore, type B MAO in the fetal cerebral cortex could be estimated to be approximately 50% according to the definition of *Johnston* (5) and *Squires* (13). The infant (3 years) cortex revealed an inhibition curve similar to that of the adult cortex. Type B enzyme in the adult cortex was found to be 80–90%. The percent inhibition of MAO by $10^{-7} M$ deprenyl is also shown for all the samples in table II.

Effect of Solubilization

The solubilized preparation as well as the crude mitochondrial fraction of the fetal brain was assayed for MAO activity with kynuramine as substrate at various concentrations of deprenyl. The result is shown in figure 2. As can be seen from the figure, the crude mitochondrial fraction of the fetal brain revealed an inhibition curve similar to that obtained from the homogenate of the fetal cerebral cortex (fig. 1), showing that approximately 50% of the activity is due to type B MAO. The solubilized enzyme was found to be much more sensitive to deprenyl, and almost all the activity was inhibited by $10^{-7} M$ deprenyl. This result indicates that the activity in the solubilized enzyme is almost of type B enzyme. In the solubilized enzyme, specific activity of MAO was 39% of that in the crude mitochondrial preparation.

Discussion

In the present study, MAO activities were measured with the cerebral cortices of 4 fetuses, 1 male child (3 years) and 4 female adults. It was clearly observed that MAO activity in the adult cerebral cortex was higher than that in the fetal one. The difference may not be ascribed to sex, since it was reported that there was no difference in MAO activity of the human hindbrain due to sex (12). When the mean values of MAO activities in the fetal cortices were compared with those in the adults (table I), the difference in the activity towards PEA is more marked than that towards 5-HT, suggesting that type B MAO increases more rapidly than type A enzyme during development. This result was confirmed by the study of the inhibitor sensitivity (fig. 1, table II). Recently, it has been shown that in the developing mouse brain MAO activities towards 5-HT and *p*-dimethylaminobenzylamine follow different postnatal developmental patterns (6). *Mantle et al.* (10) have confirmed this observation using inhibitor sensitivity, and proposed the hypothesis that at the early stage of development the brain contains type A enzyme predominantly and that type B MAO is successively induced during development or differentiation. Our result with the human cerebral cortex supports this hypothesis. However, it should be noted that in pig brain (14) and chick brain (17) the proportion of type B MAO to type A MAO does not increase during development. In order to explain the developmental difference of the two types of MAO in different animal species, further investigation is required.

By the solubilization of MAO in the fetal brain, the multiplicity of MAO as a function of deprenyl sensitivity was abolished and only type B enzyme was recovered (fig. 2). Although it was reported that conversion of type A MAO into type B MAO occurs during the treatment of MAO with sodium perchlorate, a chaotropic agent (4, 18), the presently observed abolition of multiple forms of MAO by solubilization seems to be ascribed to the inactivation of type A enzyme rather than to the conversion of type A MAO into type B MAO, since the specific activity of MAO fell down to 39% of the activity in crude mitochondrial preparation during solubilization. In this connection, it is recalled that type B MAO was exclusively recovered after delipidation of rat liver mitochondria with methyl ethyl ketone, and that no transformation of type A MAO into type B enzyme occurred as a result of the extraction (3).

Summary

The multiplicity of MAO of the human cerebral cortices at different ages was studied on substrate specificity and inhibitor sensitivity. When MAO activities towards 5-HT, kynuramine and PEA in the adult cerebral cortex were compared with those in the fetal cortex, the activity of the adult was much higher than that of the fetus. The most remarkable difference in the activity between these two cortices was found towards kynuramine, followed by that towards PEA and that towards 5-HT. Using kynuramine as substrate, it was observed that in the fetal cortex approximately 50% of the activity was inhibited by $10^{-7}M$ deprenyl, while in the infant and the adult cortex, 80–90% of the activity was inhibited by the same concentration of the inhibitor. All these data show that the ratio of type B MAO to type A MAO increases during development in the human cerebral cortex. The effect of solubilization with Triton X-100 on multiplicity of MAO was studied with the crude mitochondrial fraction of the fetal brain. It was observed that only type B MAO was recovered after its solubilization.

References

1 *Century, B. and Rupp, K.L.:* Comment on microfluorometric determination of monoamine oxidase. Biochem. Pharmac. *17:* 2012–2013 (1968).
2 *Edwards, D.J.:* Monoamine oxidases in brain and platelets: implications for role of trace amines and drug action; in *Usdin and Sandler* Trace amines and the brain, pp. 59–81 (Dekker, New York 1976).
3 *Ekstedt, B. and Oreland, L.:* Effect of lipid-depletion on the different forms of monoamine oxidase in rat liver mitochondria. Biochem. Pharmac. *25:* 119–124 (1976).
4 *Houslay, M.D. and Tipton, K.F.:* The nature of the electrophoretically separable multiple forms of rat liver monoamine oxidase. Biochem. J. *135:* 173–186 (1973).
5 *Johnston, J.P.:* Some observations upon a new inhibitor of monoamine oxidase in brain tissue. Biochem. Pharmac. *17:* 1285–1297 (1968).
6 *Jourdikian, F.; Tabakoff, B., and Alivisatos, S.G.A.:* Ontogeny of multiple forms of monoamine oxidase in mouse brain. Brain Res. *93:* 301–308 (1975).
7 *Kraml, M.:* A rapid microfluorimetric determination of monoamine oxidase. Biochem. Pharmac. *14:* 1684–1686 (1965).
8 *Layne, E.:* Spectrophotometric and turbidimetric methods for measuring proteins; in *Colowick and Kaplan* Methods in enzymology, vol. III, pp. 447–454 (Academic Press, New York 1957).
9 *Lowry, O.H.; Rosebrough, N.J.; Farr, A.L., and Randall, R.J.:* Protein measurement with the Folin phenol reagent. J. biol. Chem. *193:* 265–275 (1951).
10 *Mantle, T.J.; Garrett, N.J., and Tipton, K.F.:* The development of monoamine oxidase in rat liver and brain. FEBS Lett. *64:* 227–230 (1976).
11 *McCaman, R.E.; McCaman, M.W.; Hunt, J.M., and Smith, M.S.:* Microdetermination of monoamine oxidase and 5-hydroxytryptophan decarboxylase activities in nervous tissues. J. Neurochem. *12:* 15–23 (1965).
12 *Robinson, D.S.; Davis, J.M.; Nies, A.; Ravaris, C.L., and Sylwester, D.:* Relation of sex and aging to monoamine oxidase activity of human brain, plasma, and platelets. Archs gen. Psychiat. *24:* 536–539 (1971).
13 *Squires, R.F.:* Multiple forms of monoamine oxidase in intact mitochondria as characterized by selective inhibitors and thermal stability: a comparison of eight mammalian species; in *Costa and Sandler* Monoamine oxidases – new vistas. Advances in biochemical psychopharmacology, vol. 5, pp. 355–370 (Raven Press, New York 1972).
14 *Stanton, H.C.; Cornejo, R.A.; Mersmann, H.J.; Brown, L.J., and Mueller, R.L.:* Ontogenesis of monoamine oxidase and catechol-O-methyl transferase in various tissues of domestic swine. Archs int. Pharmacodyn. Thér. *213:* 128–144 (1975).

15 Suzuki, O.; Noguchi, E., and Yagi, K.: A simple micro-determination of type B monoamine oxidase. Biochem. Pharmac. *25:* 2759–2760 (1976).
16 Suzuki, O.; Noguchi, E., and Yagi, K.: Monoamine oxidase in developing chick retina. Brain Res. *135:* 305–313 (1977).
17 Suzuki, O. and Yagi, K.: Unpublished result.
18 Tipton, K.F.; Houslay, M.D., and Garrett, N.J.: Allotopic properties of human brain monoamine oxidase. Nature new Biol. *246:* 213–214 (1973).
19 Yang, H.-Y.T. and Neff, N.H.: β-Phenylethylamine: a specific substrate for type B monoamine oxidase of brain. J. Pharmac. exp. Ther. *187:* 365–371 (1973).
20 Yang, H.-Y.T. and Neff, N.H.: The monoamine oxidases of brain: selective inhibition with drugs and the consequences for the metabolism of the biogenic amines. J. Pharmac. exp. Ther. *189:* 733–740 (1974).
21 Youdim, M.B.H. and Holzbauer, M.: Physiological aspects of the oxidative deamination of monoamines; in *Wolstenholme and Knight* Monoamine oxidase and its inhibition, pp. 105–133 (Elsevier, Amsterdam 1976).

O. Suzuki, MD, PhD, Department of Legal Medicine, Faculty of Medicine, University of Nagoya, *Nagoya 466* (Japan)

Uptake, Storage and Transport of Neurotransmitters

Maturation of Neurotransmission. Satellite Symp., 6th Meeting Int. Soc. Neurochemistry, Saint-Vincent 1977, pp. 108–115 (Karger, Basel 1978)

Uptake of ³H-GABA in Organotypic Cultures of Fetal Human and Newborn Rat Nervous Tissue

Elisabeth Hösli and L. Hösli[1]

Department of Physiology, University of Basel, Basel

Introduction

γ-Aminobutyric acid (GABA) has been proposed to act as inhibitory transmitter substance in various parts of the mammalian CNS (1). There is considerable evidence that specific uptake mechanisms are involved in terminating the action of neurotransmitters such as monoamines and amino acids at central synapses (14, 15). Studies on the cellular localization of the uptake of neurotransmitters in CNS tissue have mainly been performed on slices and homogenates using autoradiographic techniques (3, 4, 16, 20). Since electron-microscopic investigations have demonstrated that there is a considerable damage of tissue in slices, nervous tissue cultures have proved to be an excellent tool for autoradiographic studies on the cellular localization of the uptake of neurotransmitters (6–13). In the present paper a study was made on the uptake of ³H-GABA in organotypic cultures from various regions of human and rat nervous system using autoradiography.

Material and Methods

Explants from the cerebellum, brain stem and spinal cord with or without attached dorsal root ganglia of human fetuses (9–12 weeks *in utero*) and newborn rats were grown in the Maximov assemblies for 8–28 days at 35 °C (for details, see 13). For the autoradiographic investigations the cultures were incubated for 30 sec to 10 min in Hanks' solution (37 °C) containing ³H-GABA (NEN, specific activity 5.6 Ci/mM) in a concentration of 10^{-6} M. After incubation the cultures were rinsed, fixed in 3% glutaraldehyde in 0.1 M phosphate buffer, dehydrated and mounted on object slides. The air-dried cultures were covered with Ilford L 4 emulsion by the loop technique and stored in light-tight boxes for 2

[1] We are grateful to Miss *F. Maeder* for typing the manuscript and to Mr. *M. Wymann* for photographic work.

weeks. Development of the autoradiographs was performed with Kodak D 19 developer (for details, see 13). In order to obtain information whether ^{3}H-GABA is taken up by an active transport system, some studies were carried out in sodium-free incubation medium (sodium being replaced by choline and Tris) or at a temperature of 0 °C.

Results and Discussion

Uptake of ^{3}H-GABA into Neurones
Cerebellum

Electrophysiological and biochemical studies provide evidence that Purkinje cells and cerebellar interneurones may use GABA as their transmitter (1). Furthermore, cerebellar neurones possess a specific high affinity uptake system for this amino acid (15).

After incubation of human and rat cerebellar cultures with ^{3}H-GABA, many neurones revealed a strong autoradiographic reaction. The majority of labelled neurones were multipolar and appeared to be interneurones such as stellate, basket or Golgi cells (fig. 1C). The labelling was found over the cell body and processes of these neurones; the nucleus usually revealed no autoradiographic reaction (fig. 1C). Many large cells showing morphological features of Purkinje cells were also intensely labelled (fig. 1A) (8, 19, 24), a finding which is in contrast to autoradiographic studies in the cerebellum *in vivo* or in slices (3, 4, 18, 20) where no labelling of Purkinje cells by GABA was observed (fig. 1B). It is well known that Bergman glia which tightly surround Purkinje cells *in vivo* take up GABA rapidly (3, 4), whereas this glial barrier might be disturbed or absent in cultures, and therefore cultured Purkinje cells are able to accumulate ^{3}H-GABA from the incubation medium (8, 19, 24). Investigations on the uptake of glycine — another inhibitory transmitter candidate in the mammalian CNS (1) — have shown that in contrast to ^{3}H-GABA, only few cerebellar neurones have accumulated ^{3}H-glycine; the labelling being much weaker than after incubation with GABA (fig. 1D) (8). Similar observations of differences in the uptake pattern between GABA and glycine were also made in slices of rat cerebellum (5) and after intraventricular injection of ^{3}H-glycine in the cerebellum *in vivo* (20).

Spinal Cord and Brain Stem

A great number of electrophysiological and biochemical investigations also suggest that GABA may play a transmitter role in the spinal cord and in the brain stem (1). High affinity transport mechanisms for GABA have also been found in both regions (17).

After incubation of human and rat spinal cord and brain stem cultures with ^{3}H-GABA, a relatively large number of neurones were labelled (fig. 1E, F). It was observed that in spinal cord cultures, mainly small neurones have accumulated the amino acid (9), whereas in brain stem cultures large and small neurones were heavily labelled (10, 11). There was also a large proportion of neurones which showed only a weak or no autoradiographic reaction (fig. 1E). It has been suggested that GABA is taken up by neurones utilizing this amino acid as transmitter substance (3, 4, 16, 20). *Iversen and Bloom* (16) have observed that after labelling of spinal cord homogenates with a mixture of ^{3}H-GABA and ^{3}H-glycine, approximately 50% of the synaptosomes revealed an autoradiographic reaction being the sum of the values obtained after incubation with either glycine (26%) or GABA (25%) alone, suggesting that GABA and glycine are taken up by different synaptosomal populations. From our studies in tissue culture it was, however, not possible to determine whether GABA is taken up by a specific neuronal population or by a specific cell type (9, 11, 13). The uptake of ^{3}H-GABA in CNS cultures was temperature and sodium dependent. Incubation at 0 °C or in sodium-free solution reduced the uptake of GABA considerably.

Dorsal Root Ganglia

As was observed by *Schon and Kelly* (21) in intact dorsal root ganglia (DRG), studies on the uptake of ^{3}H-GABA and L-^{3}H-glutamic acid in cultures of rat DRG have revealed that these amino acids were mainly accumulated by satellite glial cells and unmyelinated nerve fibres. In contrast to the lack of labelling of neurones in intact ganglia (21) it was found that in DRG cultures, neurones which have migrated far from the explant were also accumulating GABA (fig. 2A) and glutamate (*Hösli and Hösli*, unpublished observations). These observations suggest that satellite glial cells which are tightly wrapped around DRG neurones in intact ganglia prevent the amino acids being taken up into neuronal cell bodies (21), whereas in isolated DRG neurones in tissue

Fig. 1. A Rat cerebellar culture (17 days *in vitro*) after incubation with ^{3}H-GABA, 10^{-6} M for 5 min. The soma and dendrites of the Purkinje cell show a heavy accumulation of the amino acid; the nucleus is free of label. Bar: 20 µm. *B* Light-microscopic autoradiograph showing the grain distribution after incubation of a cerebellar slice with ^{3}H-GABA (AOAA pretreatment). Note the close correlation in localization between grain distribution basal to the Purkinje cell body (P) (arrows). × 570 (4). *C* Labelled neurones, probably stellate cells of a 17-day-old rat cerebellar culture. Incubation with ^{3}H-GABA, 10^{-6} M for 5 min. Bar: 30 µm. *D* Cerebellar neurone revealing a moderate autoradiographic reaction after incubation with ^{3}H-glycine, 10^{-6} M for 5 min. Culture 17 days *in vitro*. Bar: 20 µm (8). *E* Light-microscopic autoradiograph of a rat spinal cord culture (26 days *in vitro*) incubated with ^{3}H-GABA (10^{-6} M for 5 min). Several cell bodies are heavily labelled (arrows), whereas the majority of cells (asterisks) are covered only by few grains. × 450 (9). *F* Human brain stem culture (fetus 12 weeks *in utero*), 19 days *in vitro*. Intensely labelled neurones after incubation of the culture with ^{3}H-GABA, 10^{-6} M for 5 min. Bar: 30 µm.

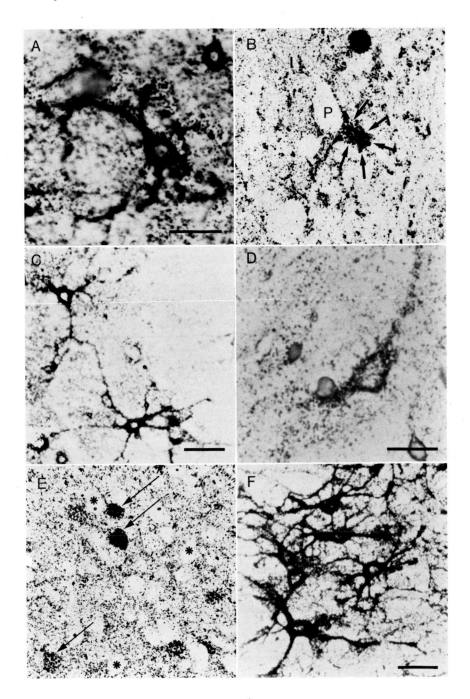

culture, this glial barrier does no more exist and therefore the amino acids can be accumulated by neurones. Similar explanations have been given for the uptake of ^{3}H-GABA in cultured Purkinje cells (see 'Cerebellum').

Uptake of Amino Acid Transmitters into Glial Cells

Uptake of amino acid transmitters into glial cells of different origin has been described by several laboratories (2, 4, 5, 11, 20, 21). As was observed with neurones, glial cells also possess a high affinity transport system for the uptake of amino acids (2, 21–23).

In cultures of human and rat CNS, uptake of ^{3}H-GABA and other amino acid transmitters such as glycine, glutamate, and asparate was also found to a great extent in glial cells (fig. 2B, C, E) (7–13). As was observed with neurones, the amino acids were distributed over the soma and processes of the glial cells, most of which appear to be protoplasmic astrocytes (fig. 2B, C, E). In contrast to the uptake into neurones, where only certain cells showed an autoradiographic reaction, almost all glial cells have accumulated the amino acids (fig. 2B). There was also a difference in time course of the uptake between neurones and glial cells. An intense labelling of glial cells was mainly observed after incubation times of 5–10 min, whereas neurones revealed already a strong autoradiographic reaction after 30 sec to 5 min incubation time (7, 8, 11). These findings are consistent with studies by *Schon and Kelly* (22) demonstrating that the rate of GABA uptake in satellite glial cells of DRG is slower than in cortical neurones, suggesting different transport systems for the uptake of amino acids into neurones and into glial cells.

In contrast to the uptake pattern of amino acid transmitters, no labelling of glial cells could be observed after incubation of brain stem and cerebellar cultures with monoamines (fig. 2D, F) (6, 8, 11). Monoamines were only taken

---

Fig. 2. A Accumulation of ^{3}H-GABA by an isolated DRG neurone in tissue culture (culture: 11 days *in vitro*, incubation with ^{3}H-GABA, 10^{-6} M for 3 min). The neurone has migrated approximately 400 μm away from the explant. Bar: 20 μm. *B* Heavily labelled glial cells in the outgrowth zone of a rat cerebellar culture (17 days *in vitro*) after incubation with ^{3}H-GABA, 10^{-6} M for 2 min. Bar: 30 μm. *C* Human spinal cord culture (fetus 12 weeks *in utero*), 18 days *in vitro* after incubation with L-^{3}H-glutamic acid, 10^{-6} M for 5 min. Intensely labelled nerve fibres and glial cells are growing out from the explant (EXPL.) of the culture. Bar: 200 μm. *D* Rat brain stem culture, 22 days *in vitro* after incubation with ^{3}H-noradrenaline, 10^{-6} M for 5 min. In contrast to figure C, only outgrowing nerve fibres but no glial cells are labelled. Bar: 100 μm. *E* Dark-field illumination picture of an intensely labelled astrocyte of a rat spinal cord culture (28 days *in vitro*). Incubation with L-^{3}H-aspartic acid, 10^{-6} M for 10 min. Bar: 20 μm. *F* Autoradiograph of a rat cerebellar culture (19 days *in vitro*) after incubation with ^{3}H-noradrenaline, 10^{-6} M for 5 min (toluidine blue staining). Intensely labelled nerve fibres are passing unlabelled neurones, most of which appear to be Purkinje cells. The glial cells (arrows) did not accumulate ^{3}H-noradrenaline. Bar: 30 μm (8).

Uptake of ³H-GABA in Nervous Tissue Cultures

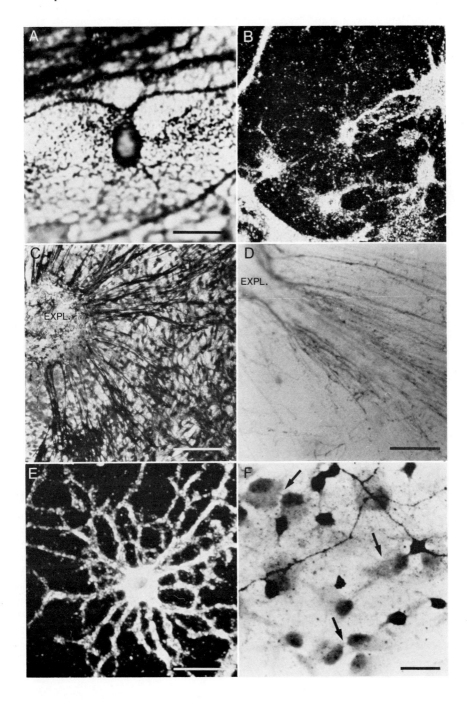

up by few neurones and by a great number of nerve fibres but not by glial elements (6, 8, 11). Figure 2C and D illustrates this difference in uptake pattern between amino acids and monoamines. After incubation with L-^{3}H-glutamic acid, outgrowing fibres, neurones, and glial cells reveal an autoradiographic reaction (fig. 2C), whereas after incubation with ^{3}H-noradrenaline mainly the outgrowing nerve fibres are labelled (fig. 2D) but no glial cells.

From our results it is concluded that glial cells might also be involved in the inactivation of amino acid transmitters and that nervous tissue cultures are a useful model for studying the cellular localization of the uptake of neurotransmitters using autoradiographic techniques.

Summary

The cellular localization of the uptake of ^{3}H-GABA was studied in cultures of cerebellum, brain stem, spinal cord, and DRG of human fetuses and newborn rats using autoradiography. In all CNS cultures, ^{3}H-GABA was found to be accumulated by a great number of neurones and to a great extent also by glial cells suggesting that glial elements might take part in the inactivation of this amino acid. In cerebellar cultures, uptake of GABA was observed into Purkinje cells and into various types of interneurones. In spinal cord and brain stem cultures, small as well as large neurones revealed a strong autoradiographic reaction, but there was also a great number of neurones which were unlabelled. In cultured DRG, ^{3}H-GABA was mainly accumulated by satellite glial cells and by some isolated DRG neurones.

References

1 *Curtis, D.R. and Johnston, G.A.R.:* Amino acid transmitters in the mammalian central nervous system; in Reviews of physiology, vol. 69, pp. 97–188 (Springer, Berlin 1974).
2 *Henn, F.A. and Hamberger, A.:* Glial cell function. Uptake of transmitter substances. Proc. natn. Acad. Sci. USA *68:* 2686–2690 (1971).
3 *Hökfelt, T. and Ljungdahl, Å.:* Autoradiographic identification of cerebral and cerebellar cortical neurons accumulating labeled gamma-aminobutyric acid (^{3}H-GABA). Expl Brain Res. *14:* 354–362 (1972).
4 *Hökfelt, T. and Ljungdahl, Å.:* Application of cytochemical techniques to the study of suspected transmitter substances in the nervous system. Adv. Biochem. Psychopharmacol. *6:* 1–36 (1972).
5 *Hökfelt, T. and Ljungdahl, Å.:* Histochemical determination of neurotransmitter distribution. Neurotrans. Res. Publ. ARNMD *50:* 1–24 (1972).
6 *Hösli, E.; Bucher, U.M., and Hösli, L.:* Uptake of ^{3}H-noradrenaline and ^{3}H-5-hydroxytryptamine in cultured rat brain stem. Experientia *31:* 354–356 (1975).
7 *Hösli, E. and Hösli, L.:* Uptake of L-glutamate and L-aspartate in neurones and glial cells of cultured human and rat spinal cord. Experientia *32:* 219–222 (1976).
8 *Hösli, E. and Hösli, L.:* Autoradiographic studies on the uptake of ^{3}H-noradrenaline and ^{3}H-GABA in cultured rat cerebellum. Expl Brain Res. *26:* 319–324 (1976).
9 *Hösli, E.; Ljungdahl, Å.; Hökfelt, T., and Hösli, L.:* Spinal cord tissue cultures. A

model for autoradiographic studies on uptake of putative neurotransmitters such as glycine and GABA. Experientia *28:* 1342–1344 (1972).
10 *Hösli, L. and Hösli, E.:* Autoradiographic localization of the uptake of glycine in cultures of rat medulla oblongata. Brain Res. *45:* 612–616 (1972).
11 *Hösli, L. and Hösli, E.:* Action and uptake of neurotransmitters in CNS tissue culture; in Reviews of physiology, vol. 81, pp. 135–188 (Springer, Berlin 1978).
12 *Hösli, L.; Hösli, E., and Andrès, P.F.:* Nervous tissue culture. A model to study action and uptake of putative neurotransmitters such as amino acids. Brain Res. *62:* 597–602 (1973).
13 *Hösli, L.; Hösli, E.; Andrès, P.F., and Wolff, J.R.:* Amino acid transmitters. Action and uptake in neurons and glial cells of human and rat CNS tissue culture; in *Santini* Golgi Centennial Symp. 'Perspectives in Neurobiology', pp. 473–488 (Raven Press, New York 1975).
14 *Iversen, L.L.:* The uptake and storage of noradrenaline in sympathetic nerves, p. 253 (Cambridge University Press, London 1967).
15 *Iversen, L.L.:* The uptake, storage, release, and metabolism of GABA in inhibitory nerves; in *Snyder* Perspectives in neuropharmacology, pp. 75–111 (Oxford University Press, New York 1972).
16 *Iversen, L.L. and Bloom, F.E.:* Studies of the uptake of ^{3}H-GABA and (^{3}H)glycine in slices of homogenates of rat brain and spinal cord by electron microscopic autoradiography. Brain Res. *41:* 131–143 (1972).
17 *Iversen, L.L. and Johnston, G.A.R.:* GABA uptake in rat central nervous system. Comparison of uptake in slices and homogenates and the effects of some inhibitors. J. Neurochem. *18:* 1939–1950 (1971).
18 *Iversen, L.L. and Kelly, J.S.:* Uptake and metabolism of γ-aminobutyric acid by neurones and glial cells. Biochem. Pharmac. *24:* 933–938 (1975).
19 *Lasher, R.S.:* The uptake of (^{3}H)GABA and differentiation of stellate neurons in cultures of dissociated postnatal rat cerebellum. Brain Res. *69:* 235–254 (1974).
20 *Schon, F. and Iversen, L.L.:* The use of autoradiographic techniques for the identification and mapping of transmitter-specific neurones in the brain. Life Sci. *15:* 157–175 (1974).
21 *Schon, F. and Kelly, J.S.:* Autoradiographic localization of (^{3}H)GABA and (^{3}H)glutamate over satellite glial cells. Brain Res. *66:* 275–288 (1974).
22 *Schon, F. and Kelly, J.S.:* The characterisation of (^{3}H)GABA uptake into the satellite glial cells of rat sensory ganglia. Brain Res. *66:* 289–300 (1974).
23 *Schrier, B.K. and Thompson, E.J.:* On the role of glial cells in the mammalian nervous system. Uptake, excretion, and metabolism of putative neurotransmitters by cultured glial tumor cells. J. biol. Chem. *249:* 1769–1780 (1974).
24 *Sotelo, C.; Privat, A., and Drian, M.-J.:* Localization of (^{3}H)GABA in tissue culture of rat cerebellum using electron microscopy radioautography. Brain Res. *45:* 302–308 (1972).

Dr. *E. Hösli*, Department of Physiology, University of Basel, *CH–4051 Basel* (Switzerland)

Neurotransmission-Related Development of the Chick Optic Lobe: Effects of Denervation, Noninnervation and Sensory Deprivation[1]

S.C. Bondy, M.E. Harrington, R. Nidess and J.L. Purdy

Departments of Neurology and Pharmacology, University of Colorado School of Medicine, Denver, Colo.

Introduction

A wide variety of biochemical changes related to neurotransmission occur during development, such as the appearance of enzymes involved in the synthesis and degradation of neurotransmitters, as well as the neurotransmitters themselves. In addition, systems related to the synaptic translocation of transmitter compounds gradually develop. These include mechanisms of synaptic neurotransmitter release, characterized by dependence of calcium and potassium ions (*Cotman et al.,* 1976) and presynaptic re-uptake of neurotransmitters or neurotransmitter precursors. This latter system is sodium and energy dependent and has the ability to concentrate compounds from very dilute solutions. These features distinguish high affinity transport systems largely confined to nerve tissue from the ubiquitous low affinity uptake processes (*Kuhar,* 1973). The high affinity binding sites, specific for neurotransmitters and their agonists and antagonists, also characterize maturing nerve tissues. These binding sites may be neuronal or glial but the use of labeled antagonists enables more specific assays of postsynaptic binding sites (*Snyder,* 1975). Since many of these phenomena are largely confined to the synapse, they may be closely related to synaptic events rather than overall brain metabolism. On the other hand, neurotransmitters and the enzymes involved in their turnover have a more broad morphological distribution. In addition, postulated amino acid neurotransmitters may also function both in intermediary metabolism and as constituents of proteins. These other properties may be quantitatively large enough to obscure the synaptic roles of amino acids. For these reasons we have chosen to concentrate

---

[1] Supported by grants from the National Institutes of Health (NS09603) and the Foundations Fund for Research in Psychiatry (70-487).

our study on the translocations of postulated neurotransmitters, rather than on their biochemical metabolism.

We are using the visual system of the chick for our studies for several reasons.

(1) The chick embryo has a well-described developmental course and is easily accessible to manipulation at all stages of development.

(2) Embryos are inexpensive to maintain and can be very uniform in their characteristics. The precise age of each embryo is known.

(3) The visual system of the chick is very dominant in its cerebral architecture and the chick uses this sensory mode to a large extent. Thus, the eyes are relatively large, making the retina easy to obtain and the optic lobes are clearly distinguishable structures. The optic nerve is a large bundle of CNS axons, making it a source of CNS tissue free of synapses and nerve cell bodies.

(4) The total crossover of the optic tract at the chiasm ensures that each eye innervates only a single optic lobe (*Cowan et al.,* 1961). This complete decussation enables a unilateral interference with visual maturation to be made, while the contralateral optic path develops normally (*Bondy and Margolis,* 1971). Such unilateral intervention by eye removal or eyelid suture can cause deficits in a single optic lobe while the other lobe in the same animal can serve as a normal control. The use of such an internal control minimizes variability so that relatively small differences between experimental and control lobes rapidly become apparent.

We have elected to intervene in tectal development in three ways.

(1) Removal of an embryonic optic cup on the third day of incubation. This prevents any fibers from contralateral retinal ganglion cells from reaching the optic tectum. The few ipsilateral connections that appear do not survive and are gone by the time of hatch (*Clarke and Cowan,* 1975). Thus, the optic tectum develops without its normal major source of afferentation. Around 95% of the sensory input to the optic lobe is by way of the retinal ganglion cell axons (*Cobb,* 1963). This lobe is referred to as 'noninnervated'.

(2) Removal of an eye after hatch. This procedure essentially denervates the contralateral lobe since the distal ends of the severed optic nerve rapidly degenerate (*Cuenod et al.,* 1970).

(3) Suture of the eyelids at hatch, thus occluding visual input to an eye. The retina receives very dim, diffuse light under these circumstances but the tectal nerve supply is completely intact. This situation is referred to as 'sensory deprivation'.

Materials and Methods

Chick and chick embryos of a White Leghorn strain were used. Eyes were removed or sutured shut under light chloroform anesthesia. Animals rapidly recovered from these procedures. After decapitation, tissues were rapidly dissected out and accurately weighed. Unilateral enucleation of 3-day-old embryos were carried out by bipolar electrocoagulation. After surgery, eggs were sealed with a coverslip held in place with paraffin wax.

High Affinity Uptake

The standard incubation medium consisted of Krebs-Ringer buffer consisting of 121 mM NaCl, 4.0 mM KCl, 1.3 mM CaCl$_2$, 1.2 mM MgSO$_4$, 1 mM ascorbic acid, 0.015 mM EDTA, 20 mM glucose and 40 mM Tris-HCl, pH 7.6. An inhibitor of monoamine oxidase (nialamide) was also present (7.5 × 10^{-5} M) as was aminoxyacetic acid (1 × 10^{-5} M), an inhibitor of γ-aminobutyric acid (GABA) transaminase. A single radioactive compound (New England Nuclear, Boston, Mass.) was then added to this medium for each incubation. Final μM concentrations of each compound were: DL-(3-^{3}H)glutamic acid, 0.01 (8.2 Ci/mmol); (2,3-^{3}H)GABA, 0.01 (10 Ci/mmol); (1,2-^{3}H)hydroxytryptamine (serotonin) 0.02 (4.25 Ci/mmol); DL-(7-^{3}H)-norepinephrine, 0.013 (7.5 Ci/mmol); (methyl-^{3}H)choline, 0.017 (2.34 Ci/mmol); (ethyl-1-^{3}H)-3,4-dihydroxyphenylethylamine (dopamine), 0.0125 (8.0 Ci/mmol).

This medium was gassed with 95% O$_2$–5% CO$_2$ and then 0.9 ml of this was mixed with 0.1 ml of either a 5% (v/v) tissue homogenate or of a synaptosomal suspension in 0.32 M sucrose. The GABA uptake mechanism was intensely active. Thus, in order to maintain linear kinetics, 0.1 ml of a 0.5% homogenate was used in these studies. In this case, 0.1 ml of 5% liver homogenate was added as a carrier at the end of the incubation. Incubation was at 37 °C for 5 min with continuous shaking. To allow for nonenergy dependent neurotransmitter binding, identical mixtures were held at 0 °C, and these served as controls.

All samples were then centrifuged at 0 °C and 28,000 g for 10 min. Supernatants were drawn off for determination of radioactivity remaining unbound to particulate matter. Pellets were resuspended in 4 ml isotonic (0.14 M) NaCl and recentrifuged (28,000 g, 10 min). The washed pellets were then dissolved in 0.5 ml tissue solubilizer. Radioactivity in these samples was determined. It was thus possible to calculate what percentage of the total radioactive compound in each incubation tube was actively taken up by the particulate fraction.

Measurement of Cerebral Blood Flow, Inulin Space and Proline Penetrance

Blood flow was measured by intracardiac injection of 0.1 ml n-methyl-^{14}C antipyrine (10 μl 20.0 mCi/mM) into 20-day-old operated embryos. After 10 sec, optic lobes were dissected out, weighed and dissolved at 50 °C in 0.5 ml NCS tissue solubilizer and assayed for radioactivity.

The inulin space was estimated after intracardiac injection of a mixture of ^{14}C-antipyrine and ^{3}H-inulin as previously described (*Purdy and Bondy,* 1976). The blood-brain barrier toward proline was estimated by a similar injection of a mixture of ^{14}C-antipyrine and ^{3}H-proline (*Purdy and Bondy,* 1976), a method first described by *Oldendorf* (1970). This double label technique compares the penetrance of the test compound into the brain with that of freely diffusible antipyrine.

Determination of RNA, DNA and Protein

Each weighed optic lobe was homogenized in 3 ml 0.3 N HClO$_4$ at 0 °C. The homogenate was centrifuged (3,000 g, 10 min) and the precipitate resuspended in a further 3 ml 0.3 N HClO$_4$. Aliquots of this suspension were taken for determination of protein by the method of *Lowry et al.* (1951) and the rest was recentrifuged. This precipitate was extracted twice with 3 ml ethanol and the final pellet was dissolved in 3 ml 0.3 N KOH and maintained at 37 °C for 2 h. After cooling, the solution was made acid with 0.12 ml 70% HClO$_4$ and centrifuged. The pellet was washed twice with 1 ml cold 0.5 N HClO$_4$ and the absorbance of the combined supernatants at A_{260} where there was a maximum was measured. Taking a concentration of hydrolyzed RNA to have an A_{260} of 30.0 (*Bondy and Roberts*, 1967), the RNA concentration in the lobe was calculated. The residual precipitate that had been washed with cold $HClO_4$ was then heated to 90 °C for 15 min in 1 ml 0.5 N $HClO_4$ and the absorption maximum at A_{268} was measured. DNA concentrations were calculated using a calf thymus DNA preparation as a standard. The absorbance at A_{268} of 1 mg hydrolyzed DNA/ml was 32.3.

Enzyme Determinations

Acetylcholinesterase was determined colorimetrically using the rate of hydrolysis of acetylthiocholine (*Ellman et al.*, 1961). Nonspecific cholinesterase was similarly determined using butyrylthiocholine iodide as the enzyme substrate. 0.1% tissue homogenates in 0.07 M Na_2POH_4 and 0.07 M Na_2POH_4 and 0.07 M KH_2PO_4, pH 8.0 were incubated at 37 °C with 0.3 mM dithiobisnitrobenzoic acid and 0.5 mM acetylthiocholine or butyrylthiocholine.

Results

Concentration of Macromolecules and Tissue Weight

After $2^1/_2-3$ weeks, the experimental conditions of denervation or noninnervation caused major developmental failure in optic lobes (table I). There was around a 27% reduction of lobe weight, protein and RNA. DNA per lobe was similarly reduced in noninnervated lobes but was unchanged in optic lobes, denervated at hatch. This was taken to indicate that noninnervation caused extensive tectal cell death or failure of proliferation while denervation did not result in significant cell death. In this latter case, the tectum may have a normal cell number but the mean size and protein and RNA content of each cell are reduced. Minor but significant losses in weight and protein content were found in optic lobes contralateral to a sutured eye.

Table I. Weight and macromolecule content of optic lobes from chicks subjected to unilateral manipulation of visual system (mg/lobe)

	Noninnervated 20-day embryos			Enucleated 21-day chicks			Sutured 21-day chicks		
	E	C	E/C	E	C	E/C	E	C	E/C
Wet weight	49.3	67.2	0.73*	66.4	89.2	0.74*	85.2	88.3	0.96*
Protein	3.91	5.38	0.73*	5.60	7.80	0.72*	7.46	7.88	0.95*
RNA	0.077	0.106	0.73*	0.111	0.162	0.69*			
DNA	0.037	0.050	0.74*	0.067	0.067	1.00			

E = Experimental lobe, contralateral to manipulated eye, C = control, * $p < 0.05$ that E/C is below unity.

Table II. Nutrient supply to optic lobes from chicks subjected to unilateral manipulation of the visual system

	Noninnervated 20-day embryos			Enucleated 21-day chicks			Sutured 21-day chicks		
	E	C	E/C	E	C	E/C	E	C	E/C
Proline penetrance (relative to antipyrene)	0.110	0.113	0.97	0.093	0.081	1.15*	0.084	0.076	1.11*
Relative blood flow	–	–	0.96	–	–	0.85*	–	–	0.84*
α-Aminoisobutyrate concentration (relative to plasma)	2.90	3.89	0.75*	3.90	3.85	1.01	3.88	3.83	1.01
Inulin space (fraction of whole tissue)	0.167	0.170	0.98	0.132	0.135	0.98	–	–	–

E, C, * as in table I. Blood flow and proline penetrance were measured in chicks 2 days after enucleation or suture.

Nutrient Supply

The blood supply and blood-brain barrier in noninnervated lobes could not be distinguished from that in control lobes (table II) but their ability to transport amino acids (as judged by concentration of α-aminoisobutyric acid) was depressed. The opposite situation was seen in 3-week enucleated or sutured chicks — the amino acid transport capacity was unaltered but the velocity of blood flow and blood-brain barrier toward proline were reduced in experimental lobes. These changes were not attributable to any alteration of the extracellular compartment of tissue (as judged by the inulin space). The blood-brain barrier appears well before hatch in the chick (*Purdy and Bondy,* 1976). Its maintenance after hatch may in part depend on nervous activity. Similarly cerebral blood flow is regulated largely by sensory input. Since the effect of suture is as drastic as that of enucleation on these parameters, trophic influences are unlikely to play a major role in their development and maintenance.

Neurotransmitter-Related Changes

Levels of acetylcholinesterase within noninnervated embryonic lobes were depressed while the concentration of butyrylcholinesterase was unchanged (table III). Thus, while butyrylcholinesterase was reduced in proportion to the lower weight of experimental lobes, acetylcholinesterase was depressed below

Table III. Cholinesterases in optic lobes from chicks enucleated after 3 days incubation or at hatch (μmoles substrate hydrolyzed/min/g wet weight)

	Noninnervated 20-day embryos			Enucleated 21-day chicks		
	E	C	E/C	E	C	E/C
Acetylcholinesterase	23.9	28.3	0.84*	39.9	43.0	0.93*
Butyrylcholinesterase	1.18	1.20	0.98	1.87	1.92	0.97

E, C * as in table I.

this value. *Marchisio* (1969) reported an even more severe depression of choline acetyltransferase in the nonafferented embryonic optic lobe. This implies that neuronal maturation may be more severely affected by the surgery than glial maturation. 3 weeks after enucleation of new hatched chicks, acetylcholinesterase levels were also reduced more than tissue weight. Since enucleation has a minor effect of the post-hatch rise in acetylcholinesterase levels, the sharp rise of this enzyme at the time of hatching (*Vernadakis and Burkhalter*, 1967) is likely to be genetically predetermined rather than due to sensory or trophic influences.

High affinity uptake of putative neurotransmitters were studied. We have previously shown this uptake to be: (a) dependent on the sodium ion, (b) relatively specific for each compound tested, (c) low in nerve bundles not containing synapses, (d) concentrated in the synaptosomal fraction, and (e) blocked by presynaptic inhibitors.

We have also found the intensity of uptake of various compounds by the chick optic lobe to be very low at 6 days of incubation and to rise progressively up to the time of hatch (*Bondy and Purdy*, 1977).

The levels of high affinity uptake of several compounds are uniformly depressed in experimental noninnervated lobes (fig. 1). Presumably, maturation of synapses for a wide variety of neurotransmitters was impaired or prevented. The neurotransmitter uptake capacity and acetylcholinesterase levels of experimental and control lobes from sutured birds were identical. Trophic rather than sensory factors are involved in the regulation of these indices of transmitter function.

The uptake capacity per lobe was unchanged for most compounds in the experimental denervated lobe (fig. 2). Thus, the density of uptake sites was actually above control values in these lobes. This may reflect the closer synaptic packing in regions where neurons and glia have failed to grow in volume. Levels of glutamate uptake were significantly depressed to a major extent in experimental lobes. Thus, glutamate may be the primary neurotransmitter of the retinal ganglion cells. Aspartate may also be a candidate for this role because

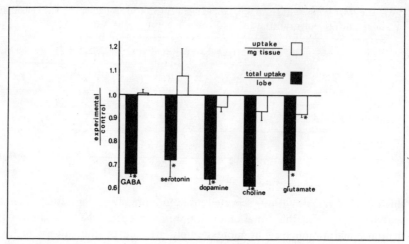

Fig. 1. High affinity uptake of compounds by homogenates of normal or nonafferented optic lobes of 21-day-old chick embryos, 18 days after removal of a single optic cup. Results are expressed as the ratio of uptake capacity of the noninnervated experimental lobe to that of the normal control lobe (open columns). The ratio of total uptake capacity of experimental to control lobes is also shown (shaded columns). Standard errors of the mean are given. * = Experimental significantly different from control ($p < 0.05$).

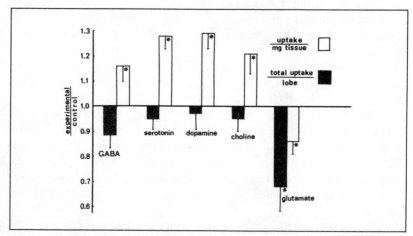

Fig. 2. High affinity uptake of compounds by homogenates of normal or denervated optic lobes of chicks 2 days after unilateral eye removal. Incubation was for 5 min in oxygenated Krebs-Ringer buffer. Results are expressed as the ratio of uptake capacity of the deafferented (experimental) lobe to that of the intact (control) lobe (open columns). The ratio of total uptake capacity of experimental to control lobes (shaded columns) is also shown. Standard errors of the mean are given. * = Experimental significantly different from control ($p < 0.05$).

aspartate and glutamate are taken up in nerve tissue by the same high affinity transport system (*Logan and Snyder,* 1971).

We have preliminary evidence, based on binding of ^{3}H-glutamate to tectal membranes, of an excess concentration of glutamate receptor sites in experimental lobes. Therefore, denervation supersensitivity may develop in the deafferented lobe.

Discussion

The distinction between genetic and environmental influence is not a clear one. The genetic program is often mediated by changes in the environment of cells. Mutual trophic interactions between cells may bring about maturation and these exchanges are often essential for differentiation of function. There is increasing evidence for a high degree of reciprocal dependence between neurons and their targets. This is maximal during maturation where cellular plasticity is high. The three experimental conditions described were chosen to contrast the effects of cerebral activity and trophic influences on brain development. Eye removal of the embryo results largely in a trophic loss to the optic tectum since the onset of electrical activity is late in embryogenesis and eggs were dark-maintained. At the other extreme, eyelid suture results in reduced sensory input but the nerve supply to the tectum is intact. Enucleation of new-hatched chicks eliminates both trophic and sensory influence from the retinal ganglion cells. We have compared the relative vulnerability of various processes to the three experimental procedures.

The results we have presented suggest that deafferentation of the optic lobe causes much less widespread disturbance of tectal neuronal organization than does prevention of formation of synapses, from retinal ganglion cell axons. While degeneration of primary synapses rapidly occurs after denervation, the synapses on many other neurons in the optic lobe appear to be present in normal numbers. Cell number is also apparently unchanged since DNA content is the same in normal and deafferented lobes. The reduced lobe weight and RNA and protein contents show that mean cell size is depressed. However, the total number of synapses per cell seems constant and neurotransmitter concentrations are not significantly altered.

The reduced development of the optic lobe of chicks unilaterally enucleated or sutured at hatch is associated with significant reductions in the rate of blood flow and glucose supply. Such changes may reflect the reduced metabolic demand of the nonfunctional lobe. However, no failure of the amino acid transport mechanism was apparent.

Removal of the optic cup of the early embryo does not alter the course of tectal development until the 13th day of incubation (*Cowan,* 1971; *Marchisio,*

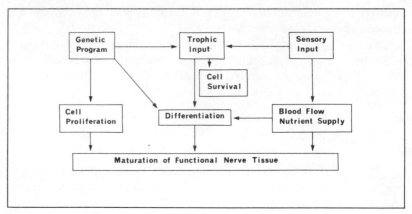

Fig. 3. Factors influencing the developmental process.

1969). But within the next week a variety of changes occur that appear to be widespread. There seems to be a loss of all types of chemical synapses studied and this may be correlated with extensive cell death suggested by the reduced DNA content of the nonafferented optic lobe. Impaired maturation of the noninnervated optic tectum is not correlatable with changes in blood flow or an altered blood-brain barrier. It is not clear whether the reduced ability of the experimental tectum to accumulate an inert amino acid is a cause or an effect of failure of adequate accretion of RNA and protein. The biochemical changes following noninnervation may be brought about by quite different mechanisms than the changes after denervation of the same region.

Figure 3 illustrates the relation of various influences to the development process. Reduced trophic input may permanently retard maturation while the later effects of sensory input changes may be more reversible. The genetic program effects cell differentiation both intrinsically and by way of cell-cell trophic interactions. The effects of sensory input may be mediated by alterations in the nutrient supply. The development and maintenance of neurotransmitter function is much more susceptible to trophic influence than to functional disuse.

Our data and that of others suggest that differentiation is intrinsic to many cerebral cells and they may have a built-in program that will take them up to a certain point of development. At a critical period, the trophic environment becomes increasingly critical for further maturation. Failure of afferentation can now lead to either arrest or regression of biochemical and morphological differentiation.

Cell death resulting from failure to make or receive appropriate synaptic connections is widespread during embryogenesis and appears to be vital for the formation of neuronal circuitry (*Clarke and Cowan*, 1975). These trophic influences are the predominant determinants in this middle period of nervous development. At later times, when the electrical circuitry of the brain begins to function, impulse patterns may modulate such intercellular chemistry.

Summary

The development of the avian optic tectum has been studied after deafferentation or sensory deprivation. Early prevention of tectal innervation (by removal of the embryonic optic cup) had severe consequences on the gross size and cell number of the noninnervated optic lobe. The maturation of high affinity neurotransmitter transport mechanisms and acetylcholinesterase levels were selectively retarded. Later denervation of the optic lobe (by unilateral enucleation of new-hatched chicks) slowed the accretion of protein and RNA leaving total tectal cell numbers unaltered. The high affinity uptake capacity of denervated lobes toward several neurotransmitters was unchanged. Thus, the maturation of tectal neurons and their synaptic contacts was more susceptible to early prevention of innervation than to later deafferentation. The ability of noninnervated optic lobes to concentrate a nonmetabolizable amino acid was impaired while the rate of blood flow in denervated lobes was reduced. Surgery at various developmental stages caused differing deficiencies in the nutrient supply of affected regions. The consequences of eyelid suture on the development of the optic lobe innervated by the deprived eye were relatively minor. Although the rate of growth was slightly depressed, no specific deficiencies of neurotransmitter-related processes were detected. A reduced rate of blood flow in the sensorily deprived optic lobe may have reflected the lower metabolic demand of the nonfunctioning region.

References

Bondy, S.C. and Margolis, F.L.: Sensory deprivation and brain development. The avian visual system as a model; in *Bures, John, Kostjuk and Pickenhain* Brain and behaviour research monograph series, vol. 4 (Fischer, Jena 1971).
Bondy, S.C. and Purdy, J.L.: Development of neurotransmitter uptake. Brain Res. *119:* 403–417 (1977).
Bondy, S.C. and Roberts, S.: Messenger ribonucleic acid of cerebral nuclei. Biochem. J. *105:* 1111–1118 (1967).
Clarke, P.G.H. and Cowan, W.M.: Ectopic neurons and aberrant connections during neural development. Proc. natn. Acad. Sci. USA *72:* 4455–4458 (1975).
Cobb, S.: Notes on the avian optic lobe. Brain *86:* 363–371 (1963).
Cotman, C.W.; Haycock, J.W., and White, W.F.: Stimulus-secretion coupling processes in brain: analysis of noradrenaline and gamma-aminobutyric acid release. J. Physiol., Lond. *254:* 475–505 (1976).
Cowan, W.M.: Studies on the development of the avian visual system; in *Pease* Cellular aspects of neural growth and differentiation, pp. 177–222 (University of California Press, Los Angeles 1971).
Cowan, W.M.; Adamson, L., and Powell, T.P.S.: An experimental study of the avian visual system. J. Anat. *95:* 545–563 (1961).
Cuenod, M.; Sandri, C., and Akert, K.: Enlarged synaptic vesicles as an early sign of

secondary degeneration in the optic nerve terminals of the pigeon. J. Cell Sci. *6:* 605–613 (1970).

Ellman, G.L.; Courtney, K.O.; Andres, V., and Featherstone, R.M.: A new and rapid colorimetric determination of acetylcholinesterase activity. Biochem. Pharmac. *7:* 88–95 (1961).

Kuhar, M.J.: Neurotransmitter uptake. A tool in identifying neurotransmitter specific pathways. Life Sci. *13:* 1623–1634 (1973).

Logan, W.J. and Snyder, S.H.: Unique high affinity uptake systems for glycine, glutamic and aspartic acids in central nervous tissue of the rat. Nature, Lond. *234:* 297–299 (1971).

Lowry, O.H.; Rosebrough, N.J.; Farr, A.L., and Randall, R.J.: Protein measurement with the Folin phenol reagent. J. biol. Chem. *193:* 265–275 (1951).

Marchisio, P.C.: Choline acetyltransferase (ChAc) activity in developing chick optic centers and the effects of monolateral removal of retina at an early embryonic stage and at hatching. J. Neurochem. *16:* 665–671 (1969).

Oldendorf, W.H.: Measurement of brain uptake of radiolabeled substances using a tritiated water internal standard. Brain Res. *24:* 372–376 (1970).

Purdy, J.L. and Bondy, S.C.: Blood-brain barrier. Selective changes during maturation. Neuroscience *1:* 125–129 (1976).

Snyder, S.H.: Amino acid neurotransmitters: biochemical pharmacology, in *Tower* The nervous system, vol. 1, pp. 355–361 (Raven Press, New York 1975).

Vernadakis, A. and Burkhalter, A.: Acetylcholinesterase activity in the optic lobes of chicks at hatching. Nature, Lond. *214:* 594–595 (1967).

S.C. Bondy, PhD, Departments of Neurology and Pharmacology, University of Colorado, School of Medicine, *Denver, CO 80262* (USA)

Axonal Transport during the Development of Retinotectal Connections in Chick Embryo[1]

P.C. Marchisio and M.F. Di Renzo

Department of Human Anatomy, University of Turin, Turin

The shape of most mature neurons is characterized by at least two types of cellular processes: the dendrites and the neurite or axon. The axon is a long and slender cytoplasmic extension which often reaches sites located at long distance from the cell body or perikaryon. A peculiar feature of the neuron metabolism is that the protein synthesizing apparatus is located around the nucleus, i.e. in the perikaryon and the large dendrites, and is virtually absent from the axon which may itself be thousandfold longer than the neuron cell body diameter and has a volume much larger than that of the whole cell body. Then, the protein synthesized in the perikaryon must migrate into the axon and move along it in order to comply with the metabolic need of the whole arborization of the axon.

Since axonal endings form synaptic connections either with other neurons or with effector organs and synapses are sites of large renewal of proteins (8), it comes clear why movements of proteins from the cell body to the periphery are critical for the correct function of neuronal networks. The function of purveying axons and synaptic endings with molecules assembled in the cell body is served by a phenomenon called proximodistal axonal transport or axonal flow.

Axonal transport has been widely reviewed (16, 25). In adult neurons, axonal transport occurs at two main velocities whose rates differ by two orders of magnitude. The rapid rate of transport (the only one being considered here) in warm-blooded animals exceeds 100 mm day^{-1}. A wide variability in the rate of rapid axonal transport has been reported: a very precise measure of its velocity has been obtained in cat motor nerves where labelled proteins have been reported to flow at a rate of $\sim$ 400 mm day^{-1} (24).

In mature neurons a variety of molecules moving along axons supports the renewal of worn-out structures both over the entire axonal length and at terminals (8). Such a continuous communication between the cell body and the

---

[1] This investigation was supported by CNR grants. P.C.M. is Professor of Histology and General Embryology of the University of Turin.

neuronal periphery is required to support synaptic transmission and the maintenance of synaptic contacts. In maturing and regenerating neurons, additional roles are most likely supported by axonal transport. Primarily, the sprouting of processes, their elongation and the formation of new branches require the continuous supply of new building blocks which cannot be provided otherwise than by axon-conveyed molecules assembled in the cell body. Moreover, the more subtle process which controls the guidance of growing axons and leads to the specific recognition of synaptic targets is endowed in surface molecules which are conveyed to growing tips by axonal flow. The nature of these phenomena, which are the basis of neuronal ontogenesis, is largely unknown: the possibility that they may be strongly dependent on and controlled by axonal transport has prompted intensive investigations on the properties of axonal flow during neuronal development.

The optic pathway of the chick embryo provides a good model for investigating the development of axonal transport. The anatomy and the embryology of the avian optic pathway are quite well known and provide good basis to follow developmental phenomena. Practical reasons also have made the chick embryonic optic system the best choice available. From early stages, the darkly pigmented and large eyeballs of the chick embryo allow easy manipulations, for example the injection of labelled precursors into the vitreous body (21). Injected precursors are taken up promptly by retinal ganglion cells wherefrom optic fibers take origin. The latter undergo complete crossing at the optic chiasm and mostly reach the contralateral optic tectum where they form synapses with the neurons of the outer tectal layers.

The initial studies on the development of axonal transport in chick embryonic optic fibers have been carried out by monitoring the migration of labelled proteins to the tectum by liquid scintillation spectrometry (21, 23). Such a migration was followed in one pathway (as compared to the opposite uninjected pathway which served as control) at different stages of development and at progressive survival times. An axonal transport of proteins was found to occur since day 10 of development: radioactive proteins reached the contralateral optic tectum in less than 2 h after intraocular administration of ^{3}H-proline. An estimate of the velocity of transport gave values in the order on 100 mm day^{-1} (23). Autoradiographic controls further confirmed the existence of an advancing front of labelled proteins in the embryonic optic pathway and suggested also that axonal transport could be revealed since day 7 of development, namely when optic axons are still actively elongating and have not yet contacted their targets in the brain (22).

From the results of our earlier experiments the concept was derived that the rate of axonal transport increased progressively with the age of the embryo, confirming data obtained elsewhere in the optic system of developing rabbits (14). In fact, the front of the wave of labelled proteins was approximately found to reach the contralateral tectum at about 2 h at any stage considered. However, the length of the optic pathway increases considerably during the same period of development and, therefore, the rate should increase comparably. It cannot be excluded that a real increase in transport rate occurs during development; a reliable measurement of the rate of rapid transport cannot be achieved in a rapidly changing system such as the embryonic optic pathway. In order to obtain information on possible development-related changes of the transport phenomenon, it was decided to investigate the 'efficiency' of transport. In other words, as a measure of the efficiency of transport, we adopted the relative amount of labelled materials which one single retina exports to the contralateral tectum in a standard time of 6 h (12). According to the latter parameter, the efficiency of axonal transport of ^{3}H-proline labelled proteins increases almost linearly between day 10 and hatching time. The increasing transport efficiency could be accounted for by an increasing rate of transport; however this seems unlikely since efficiency stops increasing immediately after hatching and then gradually decreases to reach the usual level of adult optic neurons. These results indicate that the relative amount of proteins which migrate somatofugally at high velocity is proportionally much larger in rapidly maturing retinal neurons than in those of hatched chicks where maturation proceeds at a slower rate.

Proteins transported by rapid axonal flow in developing retinal neurons are a heterogeneous mixture of polypeptides of different molecular weights (23). Among these proteins, glycoproteins represent a very interesting class in view of their well-recognized surface recognition potencies (30) and probably play an important role in the mechanism of synaptogenesis.

The axonal transport of ^{3}H-fucose labelled glycoproteins showed very interesting properties. First, their rapid transport could be very easily demonstrated in early 7-day embryos by autoradiography (13), indicating that glycoproteins represent a prominent if not the predominant component of the whole population of proteins migrating rapidly at early stages, i.e. when the axons of optic neurons are still growing towards their specific targets. In addition, it has been found that glycoproteins are very rapidly exported into nerve fibers and hence they continue to flow towards the tectum for at least 48 h finally giving rise to intense labelling of the neuropile of the tectal outer layers (13). Then, glycoproteins undergo bulk migration into axons and are later progressively released towards terminals: this peculiar phenomenon has been confirmed by blocking the efflux of glycoproteins by vinblastine (12, 19), thus adding further support to the concept that the progressive emptying of the optic fibers goes alongside with the increasing concentration of glycoproteins in the tectal neuropile (13).

An additional finding of glycoprotein transport was that their flow efficiency follows a biphasic pattern during development. After an initial rapid increase between day 13 and 15, efficiency reaches a plateau and then increases again between day 18 and hatching time; thereafter, transport efficiency decreases sharply and reaches the level of mature chicks. The biphasic development of glycoprotein transport efficiency has been confirmed for other classes of glycosylated molecules like gangliosides and glycosaminoglycans whose transport efficiency follows a developmental pattern wholly similar to that of glycoproteins (*Di Renzo and Marchisio*, in preparation, 1977). Again, this finding confirms that axonal transport of molecules mostly located in membranes is relatively more efficient in developing than in mature neurons. It also provides the information that the efficiency of their axonal transport undergoes changes related to maturational events. The earlier phase of rapid efficiency increase is in fact at least chronologically related to the bursting formation of tectal synapses following the large expansion of the arborization of retinal fibers in the tectum (6). It must also be noted that, in the same period critical changes have been recorded for acetylcholinesterase (10) and choline acetylase (17) activities and for tubulin (2) and actin (27) concentrations. The later phase of rapid increase in glycoprotein transport efficiency may well be related, at least chronologically, to the final phase of maturation of the tectal structure leading to the early recording of light-evoked reflex activity at day 18 (28).

The molecular mechanism of axonal transport is still poorly known (for review see 25). The well-known possibility of blocking axonal transport with drugs affecting the integrity of microtubules, even with different mechanism (7, 12, 15, 18) has led to believe that microtubules could represent the main structure driving axonal transport. No compelling experimental evidence has supported this hypothesis. Axonal transport must now be reexamined in the light of new available informations about other forms of intracellular motility (26). It is now largely accepted that in most cells microtubules mainly represent a rigid though rapidly remodelling skeleton which controls changes and maintenance of the cell shape. Probably microtubules may also represent anchoring structures for contractile actomyosin-containing microfilaments. Moreover, contractile microfilaments of nonmuscular cells also connect to the cytoplasmic face of the plasma membrane and thus regulate cellular motility, changes in cellular shape and the redistribution of integral molecules over the cell surface (9). There is also large evidence that intracellular movements like, for example, vesicle movements involved in secretion (1), depend on the interaction of microtubular and microfilamentous structures.

Light and electron microscopy data suggest that microtubules and microfilaments may play similar roles in nerve cells; in particular, in axons microtubules (and neurofilaments) are mostly located in the core while microfilaments occupy the marginal area underneath the axolemma (20, 31). Such a distribution which is also evident at the growth cone, has led to the hypothesis that the elongation of nervous fibers is driven by contractile structures interacting with microtubules and membrane-bound vesicles (4). In our opinion a mechanism similarly based on microtubules and microfilaments could also drive axonal transport: in such a mechanism, microtubules, being highly polarized structures, would provide mechanical support and direction to transport, while microfilaments may be directly responsible for providing the motive force. Clearly destroying microtubules impairs axonal transport, as well as axonal growth (29), because anchoring points for contractile structures become unavailable. It seems now clear that rapid axonal transport occurs within subaxolemmal channels where most transported materials have been found (3) and where most contractile structures are located (5). The characteristics of the axonal transport of glycoproteins (12) and the autoradiographic demonstration of their prevailing location at the axonal periphery (3) have led to hypothesize the additional possibility of an axolemmal movement of glycoproteins (19). The most recent knowledge about the mobility of integral proteins on the membrane plane (9) and the location of microfilaments beneath the membrane (20) seem to lend some support to this hypothesis. In maturing axons exposed glycoproteins moving on the surface of axons may dynamically interact with the surface of adjacent axons and with surrounding glial cells along the whole axonal length. This may be a basis for the parallel growth of fiber bundles and for their correct guidance towards targets.

In conclusion, axonal transport is the phenomenon driving the orderly progression of centrally synthesized molecules along the neurite. A proximodistal migration of materials is required since synthetic activity is restricted to the cell body and the use of the same materials occurs also at terminals where intense functional activity and renewal of structures take place. The amount of molecules synthesized in the perikaryon and reaching terminals is not constant but varies according to the level of synaptic activity being larger where and when more intense activity occurs. Each neuronal unit must therefore regulate the output of materials from the cell body in order to comply with the requirements of its peripheral compartment and the latter must somehow inform the cell body how large is the amount of needed materials. In our opinion, such a regulation is not accomplished by modulating the velocity of transport but may more simply take place by controlling (1) the level of perikaryal synthetic activity and (2) the size of the fraction of synthesized materials which must be shipped to the periphery. Very little is known on how such a regulation occurs. In maturing, as well as in neurons regenerating after axonal injury (11), the 'normal' level of peripherally needed materials includes also those involved in building new structures. The extra efficiency young neurons display in driving axonal transport could just be that required to comply with the high demand of structural molecules involved in the building of new axonal segments and in the formation of synapses. Therefore, control over the efficiency of axonal migration of molecules as well as the regulation of their perikaryal synthesis may be two steps whereby timing and extent of synaptogenesis may be finely regulated by embryonic neuronal populations.

Summary

An axonal transport of macromolecules has been observed and partially characterized in retinal ganglion neurons of chick embryos starting at the stage when axons are still being extended (day 7) until full maturity of the visual pathway is reached. It has been observed that the efficiency of axonal transport increases continuously with the maturational process with periods of more rapid improvement which are coincident with the onset of tectal synapses (day 13–15) and the beginning of visual activity (day 18–21). An efficient axonal transport of carbohydrate-containing macromolecules has also been studied throughout embryonic development. The mechanism underlying transport in embryonic axons depends on microtubules since mitotic poisons, like colchicine and vinblastine, block the retinotectal migration of macromolecules. However, some peculiarities of the transport in immature axons lead to suggest that some materials, notably those containing carbohydrates, may flow while associated to the axolemma. The role of axonal transport in supplying growing terminals emphasizes its importance in axonal elongation and synaptogenesis.

References

1 *Allison, A.C.:* The role of microfilaments and microtubules in cell movement, endocytosis and exocytosis; in Locomotion of tissue cells. Ciba Foundation Symposium 14, pp. 109–148 (Elsevier, North-Holland, Amsterdam 1973).
2 *Bamburg, J.R.; Shooter, E.M., and Wilson, L.:* Developmental changes in microtubule protein of chick brain. Biochemistry *12:* 1476–1482 (1973).
3 *Bennett, G.; Di Giamberardino, L.; Koenig, H.L., and Droz, B.:* Axonal migration of protein and glycoprotein to nerve endings. II. Radioautographic analysis of the renewal of glycoproteins in nerve endings of chicken ciliary ganglion after intracerebral injection of ^{3}H-fucose and ^{3}H-glucosamine. Brain Res. *60:* 129–146 (1973).
4 *Bray, D. and Bunge, M.B.:* The growth cone in neurite extension; in Locomotion of tissue cells. Ciba Foundation Symposium 14, pp. 195–209 (Elsevier, North-Holland, Amsterdam 1973).
5 *Chang, C.M. and Goldman, R.D.:* The localization of actin-like fibers in cultured neuroblastoma cells as revealed by heavy meromyosin binding. J. Cell Biol. *57:* 867–874 (1973).
6 *Cowan, W.M.:* Studies on the development of the avian visual system; in *Pease* Cellular aspects on neural growth and differentiation, pp. 177–222 (UCLA University Press, Berkeley 1971).
7 *Dahlström, A.:* Effect of colchicine on transport of amine storage granules in sympathetic nerves of rat. Eur. J. Pharmacol. *5:* 111–113 (1968).
8 *Droz, B.:* Renewal of synaptic proteins. Brain Res. *62:* 383–394 (1973).
9 *Edelman, G.M.:* Surface modulation in cell recognition and cell growth. Science *192:* 218–226 (1976).
10 *Filogamo, G.:* Recherches expérimentales sur l'activité des cholinestérases spécifique et non spécifique, dans le développement du lobe optique du poulet. Archs Biol., Liège *71:* 159–164 (1960).
11 *Grafstein, B. and Murray, M.:* Transport of protein in goldfish optic nerve during regeneration. Expl Neurol. *25:* 494–508 (1969).
12 *Gremo, F. and Marchisio, P.C.:* Dynamic properties of axonal transport of proteins and glycoproteins: a study based on the effects of metaphase blocking drugs in the developing optic pathway of chick embryos. Cell Tissue Res. *161:* 303–316 (1975).

13   Gremo, F.; Sjöstrand, J., and Marchisio, P.C.: Radioautographic analysis of ^{3}H-fucose labelled glycoproteins transported along the optic pathway of chick embryos. Cell Tissue Res. *153:* 465–476 (1974).
14   Hendrickson, A. and Cowan, W.M.: Changes in the rate of axoplasmic transport during postnatal development of the rabbit's optic nerve and tract. Expl Neurol. *30:* 403–422 (1971).
15   Kreutzberg, G.W.: Neuronal dynamics and axonal flow. IV. Blockage of intra-axonal enzyme transport by colchicine. Proc. natn. Acad. Sci. USA *62:* 722–728 (1969).
16   Lasek, R.J.: Protein transport in neurons. Int. Rev. Neurobiol. *13:* 289–324 (1970).
17   Marchisio, P.C.: Choline acetyltransferase (ChAc) activity in developing chick optic centres and the effects of monolateral removal of retina at an early embryonic stage and at hatching. J. Neurochem. *16:* 665–671 (1969).
18   Marchisio, P.C.; Aglietta, M., and Rigamonti, D.: Short-term effects of colcemid on the rapid axonal transport of proteins in the optic pathway of chick embryos. Experientia *29:* 1126–1127 (1973).
19   Marchisio, P.C.; Gremo, F., and Sjöstrand, J.: Axonal transport in embryonic neurons. The possibility of a proximo-distal axolemmal transfer of glycoproteins. Brain Res. *85:* 281–285 (1975).
20   Marchisio, P.C.; Osborn, M., and Weber, K.: Visualization of tubulin- and actin-containing structures by immunofluorescence microscopy in C-1300 neuroblastoma cells after induction of axonal sprouting; in Proc. of the ISN, vol. 6, p. 112 (1977).
21   Marchisio, P.C. and Sjöstrand, J.: Axonal transport in the avian pathway during development. Brain Res. *26:* 204–211 (1971).
22   Marchisio, P.C. and Sjöstrand, J.: Radioautographic evidence for protein transport along the optic pathway of early chick embryo. J. Neurocytol. *1:* 101–108 (1972).
23   Marchisio, P.C.; Sjöstrand, J.; Aglietta, M., and Karlsson, J.O.: The development of axonal transport of proteins and glycoproteins in the optic pathway of chick embryos. Brain Res. *63:* 273–284 (1973).
24   Ochs, S.: Characteristics and a model for fast axoplasmic transport in nerve. J. Neurobiol. *2:* 331–346 (1971).
25   Ochs, S.: Fast transport of materials in mammalian nerve fibers. Science *176:* 252–260 (1972).
26   Porter, K.R.: Motility in cells; in Goldman, Pollard and Rosenbaum Cell motility. Cold Spring Harbor Conferences on Cell Proliferation, book A, pp. 1–28 (CSH Laboratory, Cold Spring Harbor 1976).
27   Santerre, R.F. and Rich, A.: Actin accumulation in developing chick brain and other tissues. Devl Biol. *54:* 1–12 (1976).
28   Sedlacek, J.: Development of optic evoked potentials in chick embryos. Physiol. bohemoslov. *16:* 531–537 (1967).
29   Wessels, N.K.; Spooner, B.S.; Ash, J.F.; Bradley, M.O.; Luduena, M.; Taylor, E.; Wrenn, J.T., and Yamada, K.M.: Microfilaments in cellular and developmental processes. Science *171:* 135–139 (1971).
30   Winzler, R.J.: Carbohydrates in cell surfaces. Int. Rev. Cytol. *29:* 77–125 (1970).
31   Wuerker, R.B. and Kirkpatrick, J.B.: Neuronal microtubules, neurofilaments, and microfilaments. Int. Rev. Cytol. *33:* 45–75 (1972).

Prof. *P.C. Marchisio*, Istituto di Istologia ed Embriologia Generale, Corso M. D'Azeglio 52, *I–10126 Torino* (Italy)

Neurotoxic Action of Kainic Acid in the Developing Rat Striatum

J.T. Coyle, P. Campochiaro and E.D. London

Johns Hopkins University School of Medicine, Department of Pharmacology and Experimental Therapeutics, Baltimore, Md.

Introduction

An important factor regulating the differentiation of neurons may be the trophic influences exerted by either the presynaptic or the postsynaptic neurons. Such trophic influences can be uncovered by ablating the modulatory neurons and monitoring alterations in the development of the remaining neurons in a circuit. This technique has been successfully applied in the peripheral nervous system where, for example, regulation of the development of the sympathetic neurons by their presynaptic cholinergic input has been well documented (2, 3). With certain exceptions, however, lesions to the central nervous system to identify trophic influences have been less revealing because it is not usually possible to ablate discrete brain nuclei or axon bundles without damaging unrelated axons of passage and other neurons in the path of the instrument. The lack of selectivity of these lesions confounds interpretation of the specificity of their effects on developing central neurons. The discovery of neurotoxins selective for the catecholaminergic and serotonergic neurons has opened up new avenues for selectively lesioning these neuronal populations early in development. Unfortunately, similar toxins have not been available for causing selective degeneration of other types of neurons such as the cholinergic and GABAergic neurons.

Recently, we have demonstrated that stereotaxic injection into the rat striatum of nanomole quantities of kainic acid, a conformationally restricted analogue of glutamate, causes degeneration of neurons with their cell bodies in the striatum but spares axons of passage or of termination from extrinsic neurons (8). The striatal kainate lesion results in a 70% reduction in the presynaptic neurochemical markers for the cholinergic and GABAergic neurons in the striatum and a similar decrement in the markers for the GABAergic terminals innervating the substantia nigra (13, 14). In contrast, the presynaptic markers for the dopaminergic and serotonergic terminals innervating the striatum are not reduced (12). Bright field, histofluorescence and electron microscopic studies provide compelling evidence that the kainate lesion involves complete degeneration of neurons intrinsic to the striatum while sparing dopaminergic terminals and the corticofugal fibers in the striatum (7). The striking correlation between the neuroexcitatory activity and the neurotoxic potency of several glutamate analogues and kainic acid derivatives indicate that the neurotoxic action of kainic acid is mediated by excitatory glutamate receptors (15). The characteristic of the lesion are also compatible with the known localization of the glutamate receptors on neuronal cell bodies, which are vulnerable to the neurotoxin, and the absence of these receptors from axons, which are spared by the agent.

In a continuing investigation of the developmental neurochemistry of the nigrostriatal circuit (5, 6, 9), we have explored the effects of stereotaxic injection of kainic acid into the striatum of developing rats. Since the lesion causes degeneration of neurons upon which the dopaminergic terminals synapse as well as degeneration of the striatonigral GABAergic pathway which innervates the dopaminergic perikarya, it might reveal the influence of these neuronal systems on the development of the nigrostriatal dopaminergic pathway. In addition, since the neurotoxicity of kainate appears to be a receptor-mediated event, developmental studies may further clarify its mechanism of action.

Methods of Procedure

Term pregnant rats purchased from Sprague-Dawley (Madison, Wisc.) were housed in individual cages; at parturition (designated as day 1), the litters were culled of runts. Because of the size and malleability of the cranium of the immature rats, the pups could not be held in a small animal stereotaxic apparatus (David Kopf) until 21 days of age. Accordingly, molds were fabricated from plaster of Paris to stabilize the head for stereotaxic injection for each of the ages studied (7, 10 and 14 days). For stereotaxic coordinates, the pial surface and the intersection of the sagittal and the anterior coronal sutures were designated as 0; coordinates for injection of the head of the striatum were determined for each age group. After anesthesia with Equithesin (Jensen-Salisbury Laboratories), the rats received a stereotaxic injection in the left striatum of 10 nmol of kainic acid in a volume of 0.5 μl of artificial CSF titrated to pH 7.4. The rats were sacrificed 2 days and 14 days after injection and their striata were removed for biochemical assays. The contralateral uninjected striata and striata from uninjected litter mates served as controls.

The striata were homogenized in 20 vol of Tris 0.05 mM, pH 7.4 buffer containing 0.2% Triton X-100 and were assayed for the activities of tyrosine hydroxylase, choline acetyltransferase and glutamic acid decarboxylase as previously described (13). The synaptosomal high affinity uptake of [^{3}H]-glutamate in washed P$_2$ fractions was performed as described by *Coyle and Enna* (6) in the presence of 0.1 μM glutamate with a 5-min incubation at 30 °C; blanks (low-affinity uptake of glutamate) were performed in sodium-free medium. Receptor binding of [^{3}H]-kainic acid (4 Ci/mM; Amersham-Searle, Arlington, Ill.) was measured in striatal homogenates at a concentration of ligand of 50 nM by modification of the method of *Simon et al.* (16); to determine nonspecific binding, homogenates were incubated with [^{3}H]-kainic acid in the presence of 100 μM unlabeled kainic acid or 1 mM glutamate.

Results

Sequelae of Stereotaxic Injection of Kainic Acid into the Striatum of the Developing Rat

In the adult rat, injection of 10 nmol of kainate causes a 70–80% reduction in the activity of choline acetyltransferase and glutamate decarboxylase, presynaptic markers for the striatal cholinergic and GABAergic neurons; by 2 days after injection, the activity of tyrosine hydroxylase in the striatum, however, is increased nearly 90% above control (table I). Although activities of choline acetyltransferase and glutamate decarboxylase remain depressed up to 14 days

Table I. Effects of striatal kainate injection in developing striatum

Age at injection	Enzyme	Percent of uninjected control	
		2 days after injection	2 weeks after injection
7 days	choline acetyltransferase	80 ± 10	80 ± 3*
	glutamate decarboxylase	68 ± 4*	109 ± 14
	tyrosine hydroxylase	86 ± 2*	92 ± 6
10 days	choline acetyltansferase	63 ± 4+	58 ± 2+
	glutamate decarboxylase	74 ± 2+	93 ± 4
	tyrosine hydroxylase	91 ± 3	145 ± 14*
21 days	choline acetyltransferase	31 ± 3+	27 ± 8+
	glutamate decarboxylase	49 ± 4+	63 ± 4+
	tyrosine hydroxylase	116 ± 5	219 ± 21+
Adult	choline acetyltransferase	20 ± 3+	41 ± 6+
	glutamate decarboxylase	25 ± 3+	42 ± 5+
	tyrosine hydroxylase	187 ± 12+	94 ± 6

At the ages indicated, rats received a stereotaxic injection of 10 nmol of kainic acid in the left striatum. 2 and 14 days after injection, the rats were sacrificed and the injected and contralateral striata were assayed for the activities of tyrosine hydroxylase, choline acetyltransferase and glutamate decarboxylase. Results are presented in terms of percent of the activity of the enzymes (per mg tissue) in the injected as compared to the uninjected striata. Values are the mean (± SEM) from four or more preparations assayed in duplicate.
* $p < 0.05$ vs. control; + $p < 0.01$ vs. control.

after injection, the activity of tyrosine hydroxylase returns to control levels. In contrast to these effects of kainate in the adult striatum, injection of 10 nmol of kainic acid into the striatum of the 7-day-old rat causes only a modest 20–30% reduction in choline acetyltransferase and glutamate decarboxylase activities by 2 days after injection. By 2 weeks, the activity of glutamate decarboxylase has returned to normal. The activity of tyrosine hydroxylase is minimally affected in the acute and chronic state. The effects of kainate injected in 10-day-old rats are similar to that of the 7-day-old rat except that the decrement in choline acetyltransferase activity is somewhat greater. Notably, a significant rise in the activity of tyrosine hydroxylase is observed 2 weeks after the lesion. By 21 days after birth, the sensitivity of the cholinergic and GABAergic neurons in the striatum approaches that of the adult animal; however, glutamate decarboxylase activity recovers considerably by 14 days after injection as noted in the earlier ages. Although the activity of tyrosine hydroxylase is not significantly altered 2 days after injection, it is nearly doubled by 14 days after injection.

To clarify the basis for the reduced sensitivity to kainate in the immature animal, Nissl-stained sections through the forebrain containing the kainate-injected and contralateral striatum were examined. In the adult rat, there is a complete loss of neuronal perikarya in a radius of 1.5 mm around the injection site; the internal capsule fiber bundles that traverse the striatum are unaffected. In contrast, the kainate injection in the 7-day-old rat does not cause notable neuronal degeneration even along the cannula tract. While a mild but perceptible neuronal degeneration occurs in the striatum of the 10-day-old rat, the lesion in the 21-day-old rat is as complete as that which occurs in the adult.

Development of the Pre- and Postsynaptic Markers for the Striatal Glutamatergic Innervation

Since the neurotoxicity of kainate is mediated by glutamate receptors, we examined the development of the glutamatergic innervation to the striatum. As a presynaptic marker for the glutamatergic terminals, we measured the high affinity uptake of [^{3}H]-glutamate into washed P_2 fractions (fig. 1). At 4 days after birth, the synaptosomal uptake of glutamate per milligram of striatum is 23% of that of the adult and does not increase significantly during the subsequent week of postnatal development. Between 10 and 14 days after birth, synaptosomal uptake of glutamate doubles and by 21 days after birth is nearly 3-fold greater than at 10 days after birth. Thus, the major increase in the striatal synaptosomal uptake process for glutamate occurs during the second and third week after birth when sensitivity to kainate develops in the striatum.

The glutamate receptor was measured by the specific binding of [^{3}H]-kainic acid to striatal membrane preparations. The receptor has an affinity for kainate of 40 nM (K_D). L-Glutamate competes with [^{3}H]-kainate for the receptor site but has a 50-fold lower affinity, which is in agreement with electrophysiologic

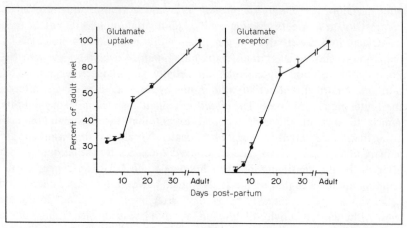

Fig. 1. Development of glutamatergic innervation to the rat striatum. Left: The uptake of [^{3}H]-glutamate (0.1 μM, 5 min, 30 °C) was measured in washed P$_2$ fractions prepared from the striata of rats of various ages. The results are presented in terms of percent of the adult uptake per milligram striatum and are the mean of three or more preparations assayed in triplicate. Right: The specific binding of [^{3}H]-kainic acid to membrane preparations from the striata of rats of various ages was measured. The results are presented in terms of percent of the adult binding per milligram striatum and are the mean of three or more preparations assayed in duplicate.

observations (1). In the adult striatum, 29 fmol of [^{3}H]-kainate are specifically bound per milligram tissue. At 4 and 7 days after birth, the levels of specific binding are quite low being respectively 2 and 6% of adult. However, over the subsequent 2 weeks of maturation, there is a dramatic 10-fold increase in receptor binding with the number of sites increasing from 1.6 to 21.6 fmol/mg tissue. By 4 weeks after birth, receptor density has achieved 80% of the adult level.

Discussion

These neurochemical and histologic studies indicate that neuronal sensitivity to the toxic action of kainic acid appears in the rat striatum at 10 days after birth and achieves adult response by 21 days after birth. In the 7-day-old rat, the agent causes only a modest reduction in choline acetyltransferase and a transient but reversible decrement in the activity of glutamate decarboxylase; neuronal degeneration is not apparent in Nissl-stained sections even along the cannula tract. In contrast, at 21 days after birth, kainic acid produces a marked and irreversible reduction in the activities of choline acetyltransferase and glutamate decarboxylase, and a complete destruction of striatal intrinsic neurons comparable to that observed in the adult rat.

The development of sensitivity to kainic acid correlates with the appearance and maturational increases in the density of glutamate receptors as monitored by the specific binding of [^{3}H]-kainic acid to striatal membranes. The potency of several glutamate analogues to compete for the receptor site with [^{3}H]-kainic acid correlates with their electrophysiologic activity as neuronal depolarizing agents (*London and Coyle,* in preparation). It should be noted, however, that there is growing evidence derived from electrophysiologic and pharmacologic studies of heterogeneity of the excitatory amino acid receptors (4); thus, [^{3}H]-kainic acid may label only one form of these receptors. But, since the ligand used is also the neurotoxin, it probably binds to the recognition site mediating the neuronal degeneration.

The striatum receives a major excitatory input from the cerebral cortex. Recent electrophysiologic and neurochemical studies indicate that these neurons use glutamate as their neurotransmitter (10, 17). With the synaptosomal high affinity uptake process for glutamate as a method for quantifying glutamatergic terminal density, there is a dramatic increase in the glutamatergic innervation to the rat striatum between 10 and 21 days after birth. It is during this time that synaptogenesis is particularly intense in the rat striatum (11). The development of the glutamatergic innervation to the striatum correlates with the development of the postsynaptic glutamate receptor as measured by sensitivity to the neurotoxic action of kainic acid or by the density of the specific binding sites for [^{3}H]-kainic acid. Thus, the development of the postsynaptic glutamate receptor is associated with the glutamatergic innervation to the striatum. In the striatum, we have previously demonstrated a temporal association between the development of the presynaptic markers for the dopaminergic terminals and dopamine sensitive adenylate cyclase (5) and development of the presynaptic markers for the cholinergic neurons and their postsynaptic muscarinic receptor (9). In contrast, *Woodward et al.* (18) observed that Purkinje cells in the immature cerebellum are sensitive to the excitatory effects of iontophoretically applied glutamate before synaptic input to the Purkinje cells is formed.

An additional goal of these studies was to examine the effects of ablation of striatal intrinsic neurons early in development on the subsequent maturation of the neurons in the nigrostriatal circuit. Although sensitivity to the neurotoxic action of kainate appears nearly a week after neuronal cell division in the striatum ceases, the major phase of neurochemical differentiation of the striatum occurs subsequently. Therefore, striatal lesions during this period might uncover trophic interactions among the neurons in the nigral striatal circuit. Notably, the effects of striatal kainate lesion at 10 and 21 days after birth on the dopaminergic terminals as monitored by tyrosine hydroxylase activity is different from that which occurs in the adult striatum. In the case of the adult, a transient increase in the activity of tyrosine hydroxylase is observed 2 days after injection; but, the activity of the enzyme returns to basal levels by 14—21 days after the lesion. The increase in striatal tyrosine hydroxylase activity in the adult is due to activation of the enzyme wherein there is a 5-fold increase in the affinity of the enzyme for its pteridine cofactor (13). In rats injected at 10 and 21 days after birth, the activity of tyrosine hydroxylase is not altered 2 days after injection; however, at 14 days after injection, the activity of tyrosine hydroxylase in the injected striatum is significantly elevated. The nature of the increase in the activity of the enzyme (e.g., activation or induction) remains to be established. A possible cause of the alteration in tyrosine hydroxylase activity may be the degeneration of the inhibitory striatonigral GABergic pathway that regulates the firing of the dopaminergic neurons (14).

An additional, interesting observation concerns the recovery of glutamate decarboxylase activity in the lesioned striatum in contrast to the persistent reduction in choline acetyltransferase activity. This recovery of glutamate decarboxylase activity suggests that surviving GABAergic fibers increase the amount of biosynthetic enzyme per fiber or that extrinsic GABAergic neurons are capable of sprouting processes to innervate the lesioned striatum. These preliminary studies indicate that kainic acid lesion of neuronal pathways in the immature brain promises to become a useful technique for examining trophic interaction among neurons in the striatum as well as in other regions of the brain.

Summary

Stereotaxic injection of the potent glutamate receptor agonist, kainic acid, into the adult rat striatum causes degeneration of neurons intrinsic to the striatum but spares axons of passage and of termination from extrinsic neurons. Sensitivity to the neurotoxic action of kainic acid as measured by alterations in the activities of choline acetyltransferase, glutamate decarboxylase and tyrosine hydroxylase and by histologic changes in Nissl-stained sections appears in the rat striatum at 10 days after birth and attains the adult response by 21 days after birth. Development of the striatal sensitivity to kainate correlates with the development of glutamatergic innervation to the striatum and of the glutamate receptor as measured by the specific binding of [^{3}H]-kainic acid.

Acknowledgement

This research was supported by USPHS grants MH 26654 and NS 13584 and grants from the National Foundation (Basil O'Connor Award) and the McKnight Foundation. J.T.C. is the recipient of an NIMH Research Career Development Award, Type II (MH 00125). P.C. is a Henry Strong Dension Research Scholar. E.L. has a USPHS fellowship (MH 07142). We thank *Vickie Rhodes* and *Barbara Spink* for secretarial assistance and *Robert Zaczek* for technical assistance.

References

1. *Biscoe, T.J.; Evans, R.H.; Headley, P.M.; Martin, M.R., and Watkins, J.C.:* Structure-activity relations of excitatory amino acids on frog and rat spinal neurons. Br. J. Pharmacol. *58:* 373–382 (1976).
2. *Black, I.B.; Bloom, E.M., and Hamill, R.W.:* Central regulation of sympathetic neuron development. Proc. natn. Acad. Sci. USA *73:* 3575–3578 (1976).
3. *Black, I.B.; Hendry, I.A., and Iversen, L.L.:* Effects of surgical decentralization and nerve growth factor on the maturation of adrenergic neurons in mouse sympathetic ganglion. J. Neurochem. *19:* 1367–1377 (1972).
4. *Buu, N.T.; Puil, E., and Gelder, N.M. van:* Receptors for amino acids in excitable tissues. Gen. Pharmacol. *7:* 5–14 (1976).
5. *Coyle, J.T. and Campochiaro, P.:* Ontogenesis of dopaminergic-cholinergic interactions in the rat striatum: a neurochemical study. J. Neurochem. *27:* 673–678 (1976).
6. *Coyle, J.T. and Enna, S.J.:* Neurochemical aspects of the ontogenesis of GABAergic neurons in the rat brain. Brain Res. *111:* 119–133 (1976).
7. *Coyle, J.T.; Molliver, M.E., and Kuhar, M.J.:* Morphologic analysis of the kainic acid lesion of rat striatum. J. comp. Neurol. (submitted, 1977).
8. *Coyle, J.T. and Schwarcz, R.:* Model for Huntington's chorea. Lesion of striatal neurons with kainic acid. Nature, Lond. *263:* 244–246 (1976).
9. *Coyle, J.T. and Yamamura, H.I.:* Neurochemical aspects of the ontogenesis of cholinergic neurons in the rat brain. Brain Res. *118:* 429–440 (1976).
10. *Fonnum, F. and Storm-Mathisen, J.:* High affinity uptake of glutamate in terminals of corticostriatal axons. Nature, Lond. *266:* 377–378 (1977).
11. *Lu, E.J. and Brown, W.J.:* The developing caudate nucleus in the euthyroid and hyperthyroid rat. J. comp. Neurol. *171:* 261–284 (1977).
12. *Schwarcz, R.; Bennett, J.P., and Coyle, J.T.:* Loss of striatal serotonin synaptic receptor binding induced by kainic acid lesion: correlates with Huntington's disease. J. Neurochem. *28:* 867–869 (1977).
13. *Schwarcz, R. and Coyle, J.T.:* Striatal lesions with kainic acid: neurochemical characteristics. Brain Res. *127:* 235–249 (1977).
14. *Schwarcz, R. and Coyle, J.T.:* Neurochemical sequelae of kainate injections in corpus striatum and substantia nigra of the rat. Life Sci. *20:* 431–436 (1977).
15. *Schwarcz, R.; Scholz, D., and Coyle, J.T.:* Structure-activity relations for the neurotoxicity of kainic acid deviations and glutamate analogues. Neuropharmacology (in press, 1977).
16. *Simon, J.R.; Contrera, J.F., and Kuhar, M.J.:* Binding of [^{3}H] kainic acid, an analogue of L-glutamate to brain membranes. J. Neurochem. *26:* 141–147 (1976).
17. *Spencer, H.J.:* Antagonism of cortical excitation of striatal neurons by glutamic acid diethylester: evidence for glutamic acid as excitatory transmitter in rat striatum. Brain Res. *102:* 91–101 (1976).
18. *Woodward, D.J.; Hoffer, B.J.; Siggins, G.R., and Bloom, F.E.:* The ontogenetic development of synaptic junctions, synaptic activation and responsiveness to neurotransmitter substances in rat cerebellar Purkinje cells. Brain Res. *34:* 73–97 (1974).

J.T. Coyle, MD, Johns Hopkins University School of Medicine, Department of Pharmacology, 725 North Wolfe Street, *Baltimore, MD 21205* (USA)

Effects of Intrinsic and Extrinsic Factors in Neuronal Maturation

Maturation of Neurotransmission. Satellite Symp., 6th Meeting Int. Soc. Neurochemistry, Saint-Vincent 1977, pp. 142–151 (Karger, Basel 1978)

Developmental and Biochemical Studies on Tubulin:Tyrosine Ligase[1]

T. Pierce, R.K. Hanson, G.G. Deanin, M.W. Gordon and A. Levi[2]

A. Ribicoff Research Center, Norwich, Conn.; Department of Biobehavioral Sciences, University of Connecticut, Storrs, Conn., and Laboratorio de Biologia Cellulare, Roma

Introduction

Some years ago, *Levi-Montalcini and Angeletti* (17) reported that the earliest observable morphological effect of nerve growth factor (NGF) is the appearance of microfilaments and microtubules in NGF-sensitive cells. Since that time, numerous studies have shown that the appearance of such cytoplasmic structures is invariably associated with cell differentiation (for review, see 28) and, in tissue culture, with changes in cell shape (4). In several systems it has been shown that the formation of microfilaments and microtubules is independent of *de novo* protein synthesis. In cultures of dorsal root ganglion cells, the presence of large amounts of cycloheximide does not prevent the appearance of microfilaments and microtubules when these cells are stimulated with NGF (33). Cultures of neuroblastoma-derived cells can be induced to sprout neurites even when there is no net synthesis of tubulin (26). Even the formation of the mitotic spindle, which contains tubulin, can occur in the absence of new protein synthesis (29). The nature of the post-translational signal which induces the formation of microfilaments and microtubules has, therefore, become a matter of intensive investigation.

The study of the conversion of tubulin into microtubules was greatly advanced when, in 1972, *Weisenberg* (32) demonstrated that tubulin can be

---

[1] This work was supported, in part, by a grant from the NIH, No. NS14240.

[2] The American authors wish to express their gratitude to Professor *Rita Levi-Montalcini* and her colleagues for their gracious hospitality and help at the Laboratorio de Biologia Cellulare where some of the work reported here was performed.

induced to form microtubular structures *in vitro*. Tubulin was shown to bind a variety of proteins which could only be removed by ion-exchange chromatography (28). These microtubular associated proteins (MAPs) include several different enzymatic activities — an ATPase, a transphosphorylase, a cyclic AMP stimulated protein kinase and other basic proteins whose biological roles are largely unknown.

That many basic proteins should form tight associations with tubulin no doubt reflects the highly acidic nature of the tubulin molecule. It is not clear, however, that these associations are critical for microtubule formation. One of these basic proteins, tau, has been shown to induce the assembly of tubulin into microtubules under conditions where assembly does not otherwise occur (31). Further, with an antibody to tau, *Connolly et al.* (5) have shown that this protein is closely associated with microtubules formed *in vivo*. Yet, tubulin can be assembled into microtubules, *in vitro*, in the absence of tau, under appropriate conditions, e.g., high Mg^{++} concentration, high concentration of glycerol (12, 14). Further, it is not known, because of the methods involved, whether the association of tau with soluble tubulin *in vivo* is as great as its association with microtubules.

Even if tau is the protein which stimulates microtubule formation *in vivo*, the mechanism for disassembly of microtubules would remain obscure. Tubulin shares with other fibrillar cytoplasmic structures the ability not only to rapidly assemble into microtubules but also to rapidly disassemble into soluble tubulin. This aspect of tubulin metabolism is as critical for the dynamics of changes in cell shape as is the formation of microtubules.

Some consideration has been given to the reversible phosphorylation of tubulin as the post-translational mechanism which regulates the alternate assembly and disassembly of microtubules (28). As mentioned earlier, one of the MAPs may be a cyclic AMP-dependent protein kinase. The evidence that reversible phosphorylation of tubulin regulates its assembly and disassembly *in vivo* is circumstantial. It is reported that in HeLa cells there is a higher phosphate content of tubulin during the S and M stages of the cell cycle than in G-1 and G-2 (28). It has also been suggested that the association of guanine nucleotides with tubulin may be important for the assembly-disassembly process (28). Tubulin has two binding sites for guanine nucleotides, one of which is relatively easily dissociated. While GTP is important for tubulin assembly *in vitro*, there is no evidence that changes in guanine nucleotide concentration correlates with assembly and disassembly *in vivo*.

In the course of investigations on the incorporation of radioactive amino acids into protein catalyzed by rat brain homogenates, another reversible post-translational modification of tubulin was discovered (2). An enzyme, tyrosine:tubulin ligase (TTL) is present in the 100,000 *g* supernatant fraction of rat brain which catalyzes the addition of tyrosine to the carboxyl terminus of the α-chain by the formation of a peptide bond (1). TTL requires ATP, Mg^{++} and K^+ for the addition of tyrosine but, in the presence of ADP and Pi, the terminal tyrosine is removed (23). While initially it was believed that this activity was exclusive to brain, it was later shown that TTL is present in all of the adult tissues of the rat, though the activity of the enzyme is higher in brain than in any other adult tissue that was examined (6). The ubiquity of this enzymatic activity is further shown by its presence in neuroblastoma (22), glioma (8) and pheochromocytoma (16) cell lines.

The activity of TTL in embryonic tissues is much greater than in comparable tissues of the adult (6). On day 14, TTL activity in the developing chick thigh muscle is 1.5 times greater than that of embryonic brain even though muscle contains only 10% of the activity of the brain in the adult. Developmental studies show a correlation between the activity of TTL and the proliferation of microtubules (7). A peak in TTL activity occurs in chick thigh muscle on developmental day 13, a time when myoblasts are being rapidly converted to myotubes (10). Myotubes contain highly ordered arrays of microtubules and their normal development is inhibited by colchicine, a drug which interferes with microtubule assembly (3). After day 13 there is a rapid drop in TTL activity but low levels of activity persist throughout adult life (7).

The developmental pattern of TTL activity in chick brain is more complex; peaks of activity occur at developmental days 10, 13 and a high point is achieved on day 17 (7). This complex developmental pattern, we thought, might be attributed to the asynchronous development of different brain areas. We therefore sought a more suitable system for the developmental study in nervous tissue and for this purpose selected the dorsal root ganglion.

The development of dorsal root ganglia in the chick has been extensively documented. There is a temporal dichotomy between proliferation and differentiation of the ventrolateral and mediodorsal cells in these ganglia, and only the latter are sensitive to NGF (9, 17). The former population differentiates early during embryogenesis. The mediodorsal cells, however, cease proliferation and begin to develop neurite extensions at day 9–10. Coincident with this process is the development of NGF sensitivity as shown by *Levi-Montalcini and Angeletti* (17), a sensitivity which was later shown to be associated with the appearance of NGF receptor sites on the plasma membrane (11). The sensitivity to NGF diminishes after day 14 when the differentiation of the mediodorsal cells appears to be essentially complete. A study of the developmental pattern of TTL activity in these ganglia, then, not only afforded an opportunity to determine if changes in TTL activity correlate in time with process formation but also with periods of sensitivity to NGF.

Since significant TTL activity is present in tubulin preparations purified by four cycles of assembly and disassembly by the method of *Shelanski et al.* (27) we also investigated the possibility that TTL itself might be one of the proteins which forms a strong association with tubulin. *Levi et al.* (15) have already shown that NGF, a basic protein, binds to tubulin with a high affinity constant. We, therefore, considered the possibility that NGF might mediate the post-translational modification of tubulin catalyzed by TTL and that a TTL-NGF-tubulin complex could somehow account for the ability of very small amounts of NGF to produce such massive changes in sensitive cells.

Materials and Methods

Developmental Studies

Fertile White Leghorn eggs were obtained from Spafas, Norwich, Conn. The eggs were incubated for various periods in a Humidaire incubator. At the time periods designated in figure 1, embryos were removed and the spinal ganglia dissected and freed from contaminating tissue, on ice. Immediately following dissection and cleaning, the ganglia were frozen at -80 °C. When samples from all age groups were collected, the ganglia were disrupted by three 15-sec bursts, at 0 °C, utilizing a Bronwill Biosonik III sonicator equipped with a micro tip. All sonicates were prepared in 200 µl of buffer A: 100 mM KCl, 50 mM MES, pH 6.6, 5 mM Mg acetate, 1 mM dithiothreitol (DTT) and 10% glycerol. The supernatant fraction was obtained by centrifugation for 1 h at 100,000 g at 0 °C, assayed for total protein content by the method of *Lowry et al.* (18), diluted to 0.6 mg/ml with buffer A and assayed for TTL activity in the presence of 3 times cycled rat brain tubulin purified by the Shelanski procedure (27). Incubation mixtures, in a total volume of 83 µl, contained 1 mg/ml tubulin, 2.5 mM ATP, 2.5 mM Mg acetate, 0.1 mg/ml RNase, 100 µM ^{3}H-L-tyrosine (specific activity 2.53 Ci/mmol), 100 mM KCl, 50 mM MES, pH 6.6, 1 mM DTT, 10% glycerol. For each developmental period, enzyme activity was measured at two concentrations of supernatant protein (0.1 and 0.2 mg/ml) to insure that substrate concentration was nonlimiting. Incubations were for 10 min at 37 °C in a Dubnoff metabolic shaker. Incorporation of labeled tyrosine into trichloracetic acid precipitable protein was measured on Whatman 3MM disks as described by *Mans and Novelli* (19) and corrected for incorporation due to enzymatic activity associated with the purified tubulin.

Biochemical Studies

Phosphocellulose (8 ml column) was equilibrated with buffer B: 50 mM MES, pH 6.6, 25 mM KCl, 0.5 mM Mg acetate, 0.5 mM ATP, 1 mM DTT and 10% glycerol, at 0–4 °C. Approximately 8 mg of protein (3.5 mg/ml in buffer B), containing tubulin purified by two cycles of assembly and disassembly plus a small amount of ^{3}H-tyrosyltubulin, was applied to the column, wahsed with 18 ml buffer B and the column developed by a linear gradient of KCl (0.025–1 M) in buffer B. Fractions (0.1 ml) were counted in a Packard Liquid Scintillation Spectrometer. TTL activity was assayed, as described above, in the presence of purified tubulin. Data was corrected for the contribution of ^{3}H-tyrosyltubulin used as marker. TTL activity was also assayed using tubulin purifed by phosphocellulose chromatography. SDS gel electrophoresis of phosphocellulose fractions was performed as described by *Laemmli* (13). Experiments on the effects of Mg^{++} and polyamines were performed under the same conditions described above except for the variation in Mg^{++} or the addition of spermine or spermidine. In experiments conducted at pH 7.4, Tris-HCl, at 50 mM replaced MES buffer. Dialyses were carried out against the appropriate incubation buffer.

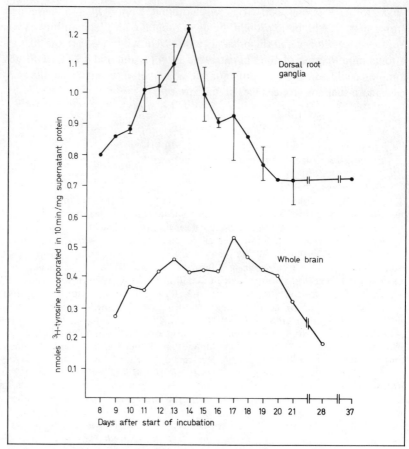

Fig. 1. Developmental pattern of TTL activity in the dorsal root ganglia and brain of the chick. In the ganglionic studies, each point represents the average of at least two separate assays; variation is indicated by crossbars which represent the highest and lowest values.

Results and Discussion

The developmental pattern of TTL activity in the dorsal root ganglion of the chick is described in figure 1 along with data from a previous experiment describing TTL activity in whole brain during embryogenesis (recalculated for 10 min incubation) (7). The activity per milligram protein is greater in the dorsal root ganglion at all stages of development. Unlike whole brain, TTL activity in the ganglion rises sharply from day 10 to day 14, at which time the activity is

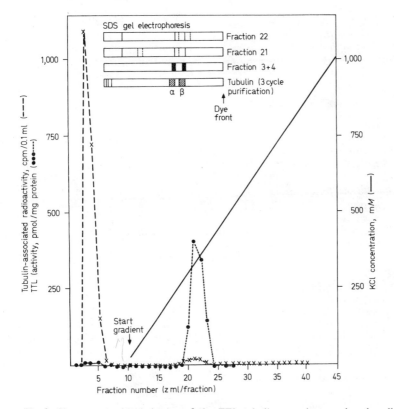

Fig. 2. Chromatographic behavior of the TTL:tubulin complex on phosphocellulose. Inserts are drawings of gels after SDS gel electrophoresis of various fractions.

some 50% greater than that occurring on day 8, the first period measured. After day 14, there is a rapid fall in TTL specific activity and by day 19, the specific activity of the enzyme is equivalent to that at day 8; TTL activity falls an additional 10% thereafter and at 38 days after the start of incubation is about the same as at hatching. The specific activity at that time, however, is almost 5 times greater than that of whole brain.

The most rapid rise in TTL specific activity corresponds with the onset of process formation by the mediodorsal cells of the ganglion and the rapid rise in NGF receptor sites (9, 11). The fall in specific activity coincides with that period when NGF binding by the spinal ganglion shows a precipitous fall, but the decrease in specific activity of the enzyme is only about 40% of the peak value. These data and those reported by *Levi et al.* (16) for an NGF sensitive pheochromocytoma cell line, where NGF added to the culture medium produces

Table I. Effect of variations in isolation procedure and assay conditions on the activity of TTL extracted from 21-day chick brain (nmoles ^{3}H-tyrosine incorporated/mg supernatant protein)

Concentration of Mg^{++}: Mg^{++}:ATP ratio:	1 mM 0.4	2 mM 0.8	2.5 mM 1.0	5 mM 2.0	10 mM 4.0
Non-dialyzed fresh enzyme					
1 Extracted at pH 6.6					
Assay at pH 6.6			0.39	0.31	0.28
2 Extracted at pH 7.4					
Assay at pH 6.6			0.37	0.31	0.30
Non-dialyzed frozen enzyme					
1 Extracted at pH 6.6					
Assay at pH 6.6	0.26	0.43			0.24
Assay at pH 7.4	0.50	0.68		0.44	0.37
2 Extracted at pH 7.4					
Assay at pH 6.6	0.26	0.41			0.21
Enzyme dialyzed at pH 6.6					
1 Extracted at pH 6.6					
Assay at pH 6.6	0.28	0.42			0.23
Assay at pH 7.4	0.45	0.65		0.43	0.38
2 Extracted at pH 7.4					
Assay at pH 6.6	0.24	0.39			0.21
Enzyme dialyzed at pH 7.4					
1 Extracted at pH 6.6					
Assay at pH 6.6	0.25	0.38			0.21
Assay at pH 7.4	0.42	0.57		0.34	0.34
2 Extracted at pH 7.4					
Assay at pH 6.6	0.23	0.37			0.21

a 40% increase in TTL activity during the first day of incubation even though the initiation of process formation is barely discernable at this time, are consistent with the following model:

NGF surface membrane interaction results in an increase in TTL activity. TTL activity somehow supports microtubule formation from a preexisting soluble tubulin pool and/or effects the stabilization of microtubular structures.

The mechanism by which TTL may stimulate microtubule formation is unknown. The ability of highly tyrosylated tubulin to form microtubules *in vitro* does not seem to differ from that of tubulin which is isolated from brain by standard methods (20). However, we do not know whether a critical amount of tyrosylated tubulin is essential for assembly, *in vitro*, since some of the

Table II. Effect of polyamines on the activity of TTL extracted from dorsal root ganglia of the 13-day chick embryo (nmoles ^{3}H-tyrosine incorporated/mg supernatant protein)

	No added polyamines (control)	Plus spermine 2 mM	Plus spermine 10 mM	Plus spermidine 2 mM	Plus spermidine 10 mM
1 Extracted at pH 6.6					
Mg^{++} 0.61 mM	0.22	0.23		0.28	
2.0 mM	0.97	0.95		1.03	
10.0 mM	0.63	0.84	0.93	0.63	0.77
2 Extracted at pH 7.4					
Mg^{++} 0.52 mM	0.13	0.18		0.21	
2.0 mM	0.84	0.90		0.94	
10.0 mM	0.58	0.75	0.81	0.63	0.73

tubulin in standard preparations contains α-chain carboxyl terminal tyrosine (21). *Thompson* (30) has reported the assembly of tubulin *in vitro* is associated with a significant loss of tyrosine and thus it seems possible that detyrosylation may play a role in microtubule formation.

The results from the phosphocellulose chromatography (fig. 2) make it clear that TTL is one of the proteins that binds to tubulin with high affinity. These data are interesting on several counts. First, phosphocellulose chromatography removes most MAPs and, in fact, is the method employed to dissociate tau protein from tubulin (31). Since some tubulin is present in the TTL peak (see also electropherograms of these fractions), after phosphocellulose chromatography and appears to be the only residual tubulin that binds to the ion exchange resin under these conditions, we must conclude that the TTL-tubulin association is very strong. Further, the net charge of this complex must differ considerably from that of free tubulin. Experiments designed to determine whether the TTL-tubulin complex has as high an affinity for NGF as does tubulin itself are underway. Clearly, studies of the enzymatic activity of such a complex and its possible role in stimulating microtubule formation *in vitro* would be critical for an evaluation of the possible role of this complex *in vivo*.

In studies of the optimal conditions required for TTL activity, we found that high concentrations of Mg^{++} were inhibitory (table I). Further, we found that both spermine and spermidine partially reversed this inhbition (table II). Since *Rubin* (24) has proposed that membrane mobilization of Mg^{++} by hormones may be important in regulating the transition from proliferation to differentiation, and since ornithine decarboxylase, which stimulates polyamine formation, is also associated with this transition (25), these observations must also be considered in evaluating the regulatory effects on differentiation which may be subserved by TTL.

Summary

TTL activity in the dorsal root ganglion of the chick increases and falls in a pattern similar to the changes in sensitivity of the mediodorsal cells to NGF. TTL has a high affinity for tubulin. The TTL activity of this complex is inhibited by high concentrations of Mg^{++}, an inhibition which is partially reversed by polyamines.

References

1 *Arce, C.A.; Barra, H.S.; Rodriguez, J.A., and Caputto, R.:* Tentative identification of the amino acid that binds tyrosine as a single unit into a soluble brain protein. FEBS Lett. *50:* 5–7 (1975).
2 *Barra, H.S.; Arce, C.A.; Rodriguez, J.A., and Caputto, R.:* Some common properties of the protein that incorporates tyrosine as a single unit and the microtubule proteins. Biochem. biophys. Res. Commun. *60:* 1384–1390 (1974).
3 *Bischoff, R. and Holtzer, H.:* The effect of mitotic inhibitors on myogenesis *in vitro.* J. Cell Biol. *36:* 111–127 (1968).
4 *Brinkley, B.R.; Fuller, G.M., and Highfield, D.P.:* Cytoplasmic microtubules in normal and transformed cells in culture. Analysis by tubulin antibody immunofluorescence. Proc. natn. Acad. Sci. USA *72:* 4981–4985 (1975).
5 *Connolly, J.A.; Kalnins, V.I.; Cleveland, D.W., and Kirschner, M.W.:* Immunofluorescent staining of cytoplasmic and spindle microtubules in mouse fibroblasts with an antibody to τ protein. Proc. natn. Acad. Sci. USA *74:* 2437–2440 (1977).
6 *Deanin, G.G. and Gordon, M.W.:* The distribution of tyrosyltubulin ligase in brain and other tissues. Biochem. biophys. Res. Commun. *71:* 676–683 (1976).
7 *Deanin, G.G.; Thompson, W.C., and Gordon, M.W.:* Tyrosyltubulin ligase activity in brain, skeletal muscle, and liver of the developing chick. Devl Biol. *57:* 230–233 (1977).
8 *Gordon, M.W.; Deanin, G.G., and Biocca, S.:* Unpublished observations.
9 *Hamburger, V. and Levi-Montalcini, R.:* Proliferation, differentiation and degeneration in the spinal ganglia of the chick embryo under normal and experimental conditions. J. exp. Zool. *111:* 457–501 (1949).
10 *Herrmann, H.; Heywood, S.M., and Marchok, A.C.:* Reconstruction of muscle development as a sequence of macromolecular syntheses. Curr. Top. devl Biol. *5:* 181–234 (1970).
11 *Herrup, K. and Shooter, E.M.:* Properties of the β-nerve growth factor receptor in development. J. Cell Biol. *67:* 118–125 (1975).
12 *Herzog, W. and Weber, K.:* In vitro assembly of pure tubulin into microtubules in the absence of microtubule-associated proteins and glycerol. Proc. natn. Acad. Sci. USA *74:* 1860–1864 (1977).
13 *Laemmli, U.K.:* Cleavage of structural proteins during the assembly of the head of bacteriophage T4. Nature, Lond. *227:* 680–685 (1970).

14 Lee, J.C. and Timasheff, S.N.: The reconstitution of microtubules from purified calf brain tubulin. Biochemistry *14:* 5183–5187 (1975).
15 Levi, A.; Cimino, M.; Mercanti, D.; Chen, J.S., and Calissano, P.: Interaction of nerve growth factor with tubulin. Studies on binding and induced polymerization. Biochim. biophys. Acta *399:* 50–60 (1975).
16 Levi, A.; Castellani, L., and Calissano, P.: Personal commun.
17 Levi-Montalcini, R. and Angeletti, P.U.: Nerve growth factor. Physiol. Rev. *48:* 534–569 (1968).
18 Lowry, O.H.; Rosebrough, N.J.; Farr, A.L., and Randall, R.J.: Protein measurement with the Folin phenol reagent. J. biol. Chem. *193:* 265–275 (1951).
19 Mans, R.J. and Novelli, G.P.: Measurement of the incorporation of radioactive amino acids into protein by a filter-paper disk method. Archs Biochem. Biophys. *94:* 48–53 (1961).
20 Raybin, D. and Flavin, M.: An enzyme tyrosylating α-tubulin and its role in microtubular assembly. Biochem. biophys. Res. Commun. *65:* 1088–1095 (1975).
21 Raybin, D. and Flavin, M.: Enzyme which specifically adds tyrosine to the α-chain of tubulin. Biochemistry *16:* 2189–2194 (1977).
22 Raybin, D. and Flavin, M.: Modification of tubulin by tyrosylation in cells and extracts and its effect on assembly *in vitro*. J. Cell Biol. *73:* 492–504 (1977).
23 Rodriguez, J.A.; Arce, C.A.; Barra, H.S., and Caputto, R.: Release of tyrosine incorporated as a single unit into rat brain protein. Biochem. biophys. Res. Commun. *54:* 335–340 (1973).
24 Rubin, H.: Magnesium deprivation reproduces the coordinate effects of serum removal or cortisol addition on transport and metabolism in chick embryo fibroblasts. J. cell. Physiol. *89:* 613–621 (1976).
25 Russell, D.H.: Polyamines in normal and neoplastic growth (Raven Press, New York 1973).
26 Seeds, N.W.; Gilman, A.G.; Amano, T., and Nirenberg, M.W.: Regulation of axon formation by clonal lines of a neural tumor. Proc. natn. Acad. Sci. USA *66:* 160–167 (1970).
27 Shelanski, M.L.; Gaskin, F., and Cantor, C.R.: Microtubule assembly in the absence of added nucleotides. Proc. natn. Acad. Sci. USA *70:* 765–768 (1973).
28 Snyder, J.A. and McIntosh, J.R.: Biochemistry and physiology of microtubules. A. Rev. Biochem. *45:* 699–720 (1976).
29 Stephens, R.E.: A thermodynamic analysis of mitotic spindle equilibrium at active metaphase. J. Cell Biol. *57:* 133–147 (1973).
30 Thompson, W.C.: Post-translational addition of tyrosine to alpha-tubulin *in vivo* in intact brain and in myogenic cells in culture. FEBS Lett. *80:* 9–13 (1977).
31 Weingarten, M.D.; Lockwood, A.H.; Hwo, S.-Y., and Kirschner, M.W.: A protein factor essential for microtubule assembly. Proc. natn. Acad. Sci. USA *72:* 1858–1862 (1975).
32 Weisenberg, R.C.: Microtubule formation *in vitro* in solutions containing low calcium concentrations. Science *177:* 1104–1105 (1972).
33 Yamada, K.M.; Spooner, B.S., and Wessells, N.K.: Ultrastructure and function of growth cones and axons of cultured nerve cells. J. Cell Biol. *49:* 614–635 (1971).

M. Gordon, PhD, A. Ribicoff Research Center, Norwich Hospital, PO Box 508, *Norwich, CT 06360* (USA)

Trophic Effects of Axonally Transported Proteins on Muscle Cells in Cultures[1]

Barry Festoff, Myron J. Duell and Hugo L. Fernandez[2]

Department of Neurology, University of Kansas Medical Center, Kansas City, Kans. and Neurobiology Research Laboratory, Veterans Administration Hospital, Kansas City, Mo.

Introduction

An extract of rat sciatic nerve, termed soluble nerve protein (SNP), had previously been shown to cause weakness and neuromuscular fatigue in SNP-immunized sheep which produced antibodies against three polypeptide components of SNP (8). Preliminary findings have also indicated that SNP appears to enhance growth and differentiation of chick embryo muscle cells in culture, an effect that is blocked by anti-SNP (6). A simplified diagram summarizing these results is shown in figure 1. It is apparent that the precise mechanism(s) underlying this distinctly neurogenic form of neuromuscular abnormality is not known. Nevertheless, these preliminary studies suggest that some component(s) of SNP serves a 'trophic' function and that the induced paralytic syndrome might involve an 'autoimmune' response to some similar endogenous factor present in the sheep. The central point of this interpretation, when taken as a working hypothesis, is the implication that the 'trophic' regulation of membrane and metabolic features of muscle are partially mediated by soluble diffusible substances of neural origin (20). Such regulatory substances, encoding 'trophic' information, might be conveyed by axonal transport (AT), for this process is known to be an important mediator in the transsynaptic control of several functional features of skeletal muscles (10).

Thus far, the biochemical nature, precise cellular origin, and mechanism(s) of action of nerve extracts which appear to exert 'trophic' influences on muscle cells in culture, are largely unanswered questions (20). It must also be noted that to date it has not been possible to identify and characterize any 'trophic'

[1] Supported in part by the Veterans Administration Medical Research Service and by a grant from the Muscular Dystrophy Association, Inc.

[2] Dr. *T.G. White,* Immunology Core Support Laboratory, VA Hospital, Kansas City, Mo., assisted with immunological studies.

Axonal Transport of Trophic Proteins

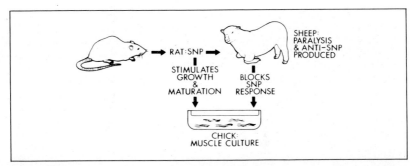

Fig. 1. Diagrammatic summary of experiments implicating the interspecies role of SNP in growth and maintenance of skeletal muscle. Rat sciatic SNP (1) induced a paralytic syndrome of sheep, (2) elicited specific antibody production in sheep and (3) stimulated growth and differentiation of chick muscle culture.

substance conveyed by AT (20). The present work was undertaken to further investigate the possible 'trophic' capabilities of SNP and to determine whether the components of SNP are conveyed by AT. To accomplish these aims we have: (a) examined the morphology and ^{125}I-α-bungarotoxin (^{125}I-αBTx) binding of muscle cultures with and without added SNP, and (b) assessed, by specific radial immunodiffusion with anti-SNP, whether SNP accumulated proximally or distally to a ligature placed around the sciatic nerve of the rat.

Methods and Results

Effects of SNP on Growth and Development of Muscle Cells in Culture

Thigh muscles of 12-day-old chick embryos were dissociated and 2×10^6 cells were plated on a collagen-coated plastic dish. The cultures were maintained in 3 ml of nutrient media as previously described (7, 16). The nutrient medium consisted of 63% Dulbecco's modified Eagle's solution; 2.5% horse serum (Millipore filtered), 2% glucose, and either 50 μl Hanks' balanced salt solution or 50 μl (100 μg protein) SNP extract every other day. In contrast to the commonly used nutrient medium (16), the medium used here was devoid of chick embryo extract and contained only 2.5%, instead of 10%, horse serum, i.e., the cultures were maintained under 'starved' conditions. The SNP extract was prepared from sciatic nerves of 150- to 200-gram male Sprague-Dawley rats according to previous methods (8). Nerves were subsequently homogenized using a Polytron PT-10 at setting 6 for 30 sec × 3 times and centrifuged at 5,000 g for 15 min. The supernatant was centrifuged at 35,000 g for 30 min and again at 105,000 g for 1 h. The resultant supernatant represented the final SNP extract.

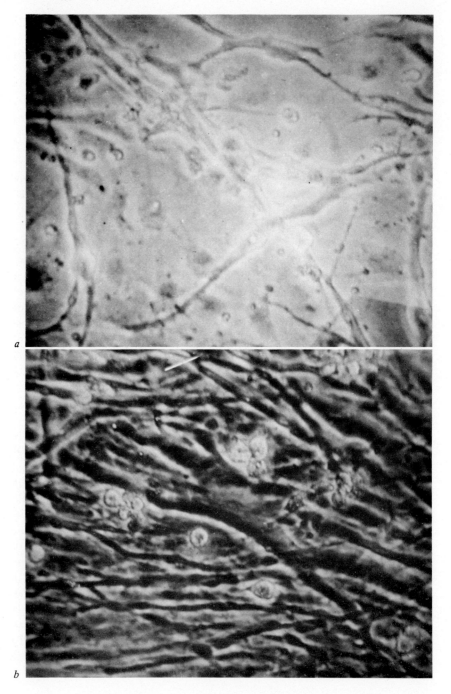

a

b

Figure 2 is a representative example of the effect of SNP on both the number and size of chick embryo myotubes formed at 6 days of culture. Note that in figure 2a, although fusion and myotube formation occurred, poorly developed, sparse populations of myotubes resulted due to the 'starvation' conditions. Figure 2b, however, shows a large number of adult-appearing, large diameter, myofibers which contracted spontaneously. Extracts of rat liver and bovine serum albumin were inactive in producing effects similar to those of SNP. Estimates of DNA and protein in the cultures correlated well with the morphological data.

Effects of SNP on Transmission-Related Proteins of Muscle Cells in Culture

Since the results described above might well have resulted from a non-specific stimulation of cellular division of both myoblasts and fibroblasts in culture, we estimated the effects of SNP on a definable muscle protein. As we were also interested in transmission-related molecules, we chose to evaluate binding of ^{125}I-αBTx. A modification (5) of the DEAE-filter disc assay (19) was used and ^{125}I-αBTx binding presumably to, or adjacent to, the nicotinic cholinergic receptor (AChR) was estimated in the presence and absence of SNP. It is important to note that neither prefusional and mitotically active myoblasts nor fibroblasts bind ^{125}I-αBTx to their surface (11, 18, 21). This was confirmed in our studies. Figure 3 represents the specific activity (pmol/mg protein) of ^{125}I-αBTx binding with and without SNP. The results indicate that: (a) SNP significantly shortened the time necessary for maximal binding to fused cells and myotubes (approximately 24 h); (b) the maximum binding in the presence of SNP was increased approximately twofold, and (c) the normal fall-off of toxin binding to contracting myofibers was seen in control cultures but, unexpectedly, the slope of this part of the curve was less steep in SNP-treated cultures. This latter observation suggests that SNP may have acted in some way to 'stabilize' the receptor on the surface of these myofibers. An analogous situation was recently reported for 'stabilization' of the acetylcholine receptor by cold α-BTx (9). Although analogous, the mechanism of such 'stabilization' must be different since enhanced toxin-binding occurred in the presence of SNP (fig. 3). In addition to these effects on acetylcholine receptors, SNP-treated cultures had significantly more acetylcholinesterase activity than control cultures (not shown).

Fig. 2. Phase contrast photomicrographs of living, 6-day-old chick muscle cultures. Polaroid positive-negative film was used to obtain prints. × 130. *a* Sparse population of small diameter myotubes grown under 'starvation' conditions (see 'Methods and Results'). *b* Adult-appearing, large diameter myofibers in large numbers in cultures grown under 'starvation' conditions plus addition of 100 μg SNP every other day. Myofibers twitched spontaneously. Magnification (as seen by size of mononucleated cells) identical to figure 2a although darker contrast.

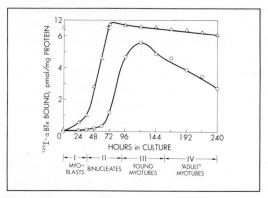

Fig. 3. Effect of SNP on a transmission-related integral muscle membrane protein. Arbitrary separation of developmental course into myoblasts, fused cells, myotubes and adult-like muscle. ^{125}I-αBTx binding was used to indicate effects of SNP on AChR. Note that (1) time to maximum binding is shortened, (2) peak of binding is increased 2-fold and (3) loss of toxin-binding from surface of cells is slowed in presence of SNP (△) compared to control (○) cultures. See 'Materials and Results' for details.

Cellular Origin and AT of SNP

Since highly specific antibodies to SNP were produced in the sheep, this afforded a unique opportunity to investigate the transport of specific SNP components. Rat sciatic nerves were ligated 2 cm distally to the sciatic notch. At various time points thereafter nerves were dissected and several 1-cm sections were cut proximally and distally to the ligature. SNP was extracted from these segments as described above and quantified by radial immunodiffusion (15) using DEAE-purified sheep anti-SNP. The results showed an average accumulation proximal to the ligature of 153% in as early as 6 h, and distal accumulation of 141% by 20 h.

This suggested that SNP was transported in retrograde as well as anterograde directions. However, the possibility existed that SNP was produced locally in either Schwann or other extra-axonal cells. To evaluate this, double ligatures were placed 1 cm apart and nerve segments were obtained. SNP accumulated at both ligatures but not between them. A representative experiment is shown in figure 4. If local production of SNP accounted for proximal or distal accumulation we would have expected to find an increase in SNP concentration between the ligatures. The level of SNP was actually less at that site, suggesting an increase in degradation as a result of sequestering this axonal protein(s). An attempt was made to determine the rate of AT of SNP by injecting ^{3}H-leucine into rat L_5 ventral spinal cord segments and measuring trichloroacetic acid-precipitable radioactivity in 2-mm sections of unilateral sciatic nerves (13) and

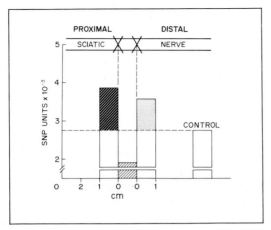

Fig. 4. Representative experiment implying AT of SNP. Double ligatures placed 1 cm apart on rat sciatic nerves. After 20 h nerves dissected and divided into 1-cm segments (proximal 0–1, 1–2 and distal 0–1 as well as interligature 0–0). SNP prepared as previously described (8) and specific radial immunodiffusion (15) used to identify SNP with sheep anti-SNP antibody (8). SNP units are arbitrary and data show that approximately 150% accumulate proximally to proximal ligature and 143% distally to distal ligature while only 60% of control SNP concentration was found between ligatures. Control (100%) is SNP content in 1-cm segment of contralateral (not ligated) sciatic nerve.

specific radial immunodiffusion of SNP in contralateral nerves. Preliminary results (not shown) indicate that SNP is transported at an intermediate or slow rate.

Discussion

The present study confirmed and extended our previous findings that SNP enhances growth and differentiation of muscle cells in culture (6). This suggested that some component(s) of SNP may have a 'trophic' role in the regulation of muscle properties, conceivably in controlling the metabolism of transmission-related proteins such as the acetylcholine receptor itself (as suggested by the syndrome produced in the sheep). Consistent with this view are experiments showing that brain, sensory ganglia, and spinal cord extracts retard the decrease in acetylcholinesterase that follows denervation of muscles in organ culture (14). Similarly, a 300,000 dalton brain extract (16) and a 10,000–50,000 dalton sciatic nerve extract (17), have been shown to stimulate morphological development, DNA and protein synthesis, and acetylcholinesterase activity, in embryonic muscle cells in culture. It has also been indicated that a 400,000–500,000 dalton brain-spinal cord extract enhances development and regulates cyclic nucleotide levels in cultured skeletal muscle cells (7).

The experiments described above leave a number of unanswered questions. For example: (a) are such hypothetical 'trophic' factors of neural origin? and (b) does in vivo 'trophic' regulation occur dependent on conveyance of such materials from their site of synthesis, presumably the cell soma, to their site of action on the target cell? We have now demonstrated that SNP, an extract that seems to have trophic capabilities, is axonally transported at an intermediate or slow rate. This latter finding also implicates the neuronal (axoplasmic) origin of SNP and indicates the need for a more detailed characterization of its transport and possible release. In this regard, it must be emphasized that AT plays a role in the 'trophic' regulation of certain electrogenic properties of muscle (1, 2), end-plate acetylcholinesterase (3), enzymatic activities (12), and protein metabolism (4).

We have begun preliminary characterization of the components of SNP. The predominant polypeptide is an acidic 65,000 ± 2,000 daltons protein which reacts with the sheep anti-SNP antibody. This protein is distinct from rat albumin, tubulin, and brain filarin. Purification of this protein is currently in progress. Whether it or some other neuronal component in SNP is the active 'trophic' factor awaits future experiments. However, the potentials for understanding the trophic functions of the neuron at a molecular level are certainly present within such studies.

Summary

A soluble protein nerve extract (SNP) is shown to have trophic capabilities. Its addition to 'starved' chick embryo muscle cultures caused enhanced growth and differentiation of myoblasts. To evaluate effects of SNP on a transmission-related macromolecule, the acetylcholine receptor, ^{125}I-αbungarotoxin binding was studied in treated and control cultures. More rapid and increased toxin-binding occurred while the fall-off of binding seen in control cultures was delayed, suggesting an effect of SNP on stabilizing the receptor. To determine what role SNP might have in vivo axoplasmic transport studies were performed using anti-SNP from sheep paralyzed in response to SNP. In vivo injection of ^{3}H-leucine and specific radial immunodiffusion demonstrated that SNP is an axoplasmic component, its transport is bidirectional, but at an intermediate or slow rate toward the terminal. Further studies are in progress to completely characterize and purify the active component.

References

1 *Albuquerque, E.X.; Warnick, J.E.; Tasse, J.R., and Sansone, R.M.:* Effects of vinblastine and colchicine on neural regulation of the fast and slow skeletal muscles of the rat. Expl Neurol. *37:* 607–634 (1972).

2 Fernandez, H.L. and Ramirez, B.U.: Muscle fibrillation induced by blockage of axoplasmic transport in motor nerves. Brain Res. 79: 385–395 (1974).
3 Fernandez, H.L. and Inestrosa, N.C.: Role of axoplasmic transport in neurotrophic regulation of muscle end-plate acetylcholinesterase. Nature, Lond. 262: 55–57 (1976).
4 Fernandez, H.L. and Ramirez, B.U.: Neurotrophic effects. Axoplasmic transport involvement in the regulation of skeletal muscle soluble proteins. Neurosci. Lett. 2: 211–216 (1976).
5 Festoff, B.W. and Engel, W.K.: In vitro analysis of the general properties and junctional receptor characteristics of skeletal muscle membranes. Isolation, purification, and partial characterization of sarcolemmal fragments. Proc. natn. Acad. Sci. USA 71: 2435–2439 (1974).
6 Festoff, B.W. and Israel, R.S.: Studies of a nerve extract with trophic properties. Neurosci. Abstr. 2: 1040 (1976).
7 Festoff, B.W. and Oh, T.H.: Neurotrophic control of cyclic nucleotide levels during muscle differentiation in cell culture. J. Neurobiol. 8: 57–65 (1977).
8 Festoff, B.W.; Israel, R.S.; Engel, W.K., and Rosenbaum, R.B.: Neuromuscular blockade with antiaxoplasmic antibodies. Neurology 27: 963–970 (1977).
9 Gardner, J.M. and Fambrough, D.M.: Properties of acetylcholine receptor turnover in cultured embryonic muscle cells (this volume).
10 Guth, L.: Trophic effects of vertebrate neurons. Neurosci. Res. Prog. Bull. 7: 1–73 (1969).
11 Hartzell, H.C. and Fambrough, D.M.: Acetylcholine receptor production and incorporation into membranes of developing muscle fibers. Devl Biol. 30: 153–165 (1973).
12 Inestrosa, N.C. and Fernandez, H.L.: Muscle enzymatic changes induced by blockade of axoplasmic transport. J. Neurophysiol. 39: 1236–1245 (1976).
13 Lasek, R.J.: Axoplasmic transport of labeled proteins in rat ventral motorneurons. Expl Neurol. 21: 41–51 (1968).
14 Lentz, T.L.: Nerve trophic function. In vitro assay of effects of nerve tissue on muscle cholinesterase activity. Science 171: 187–189 (1977).
15 Mancini, D.; Carbonara, A.O., and Theremans, J.F.: Immunochemical quantitation of antigens by single radial immunodiffusion. Immunochemistry 2: 235–254 (1945).
16 Oh, T.H.: Neurotrophic effects: characterization of the nerve extract that stimulates muscle development in culture. Expl Neurol. 46: 432–438 (1975).
17 Oh, T.H.: Neurotrophic effects of sciatic nerve extracts on muscle development in culture. Expl Neurol. 50: 376–386 (1976).
18 Prives, M.J. and Patterson, B.M.: Differentiation of cell membranes in cultures of embryonic chick breast muscle. Proc. natn. Acad. Sci. USA 71: 3208–3211 (1974).
19 Schmidt, J. and Raftery, M.A.: A simple assay for the study of solubilized acetylcholine receptor. Analyt. Biochem. 52: 349–354 (1973).
20 Smith, B.H. and Kreutzberg, G.W.: Neuron-target cellular interactions. Neurosci. Res. Prog. Bull. 14: 361–363 (1976).
21 Sytkowski, A.J.; Vogel, Z., and Nirenberg, M.W.: Development of acetylcholine receptor clusters on cultured muscle cells. Proc. natn. Acad. Sci. USA 70: 270–274 (1973).

Prof. B.W. Festoff, MD, University of Kansas Medical Center, Department of Neurology, Neurology Service, Veterans Hospital, 4801 Linwood Boulevard, Kansas City, MO 64128 (USA)

Neural Tissue Culture: A Model for the Study of the Maturation of Neurotransmission

Antonia Vernadakis[1], *Ellen B. Arnold*[2] *and Douglas W. Hoffman*[2]

University of Colorado School of Medicine, Departments of Psychiatry and Pharmacology, Denver, Colo.

Introduction

Neural tissue and cell culture are useful tools with which to study neural growth *in vitro* and to investigate factors governing cellular and molecular differentiation. Several neural culture models are available for studies of neural growth, and have been recently reviewed (17). These include organ and organotypic cultures of embryonic neural tissue, cultures of neoplastic neural cells, and cultures prepared from dissociated embryonic brain tissue. In this paper we will present data on the development of neural enzymes and on the development of norepinephrine uptake in dissociated chick embryo brain cell cultures. These data are compared with data obtained from chick embryos *in vivo* and from chick embryo brain tissue *in vitro*.

Studies in Chick Embryonic Brain

Using the chick embryo as an experimental animal model, the maturational profiles of choline acetyltransferase (ChA), the synthesizing enzyme for acetylcholine (15), and tyrosine hydroxylase (TH), the synthesizing enzyme for catecholamines (20) were examined *in vivo*. ChA activity was measured by the method of *Fellman* (4), as modified by *Weiner and Waymire* (personal commun.) and TH activity was measured by the method of *Waymire et al.* (19). The activity of ChA increased progressively with embryonic age in all brain regions

---

[1] Recipient of a Research Scientist Development Award, K02 MH42479, from the National Institute of Mental Health, National Institutes of Health.
[2] Fellow; USPHS Training Grant T 32 HD07072 from National Institute of Child Health and Human Development, National Institutes of Health.

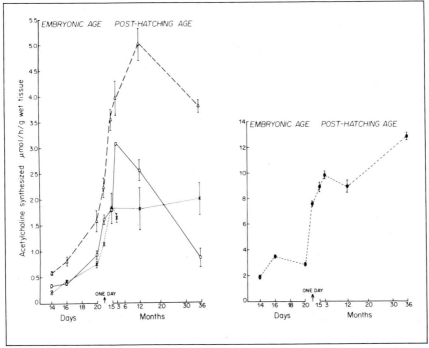

Fig. 1. Changes in ChA activity during embryonic age (days) and post-hatching age up to 36 months, in four CNS areas. ○ = Cerebral hemispheres; x = cerebellum; ● = optic lobes; △ = diencephalon/midbrain. Points with vertical lines represent means and SEM (15).

studied (fig. 1). In the cerebral hemispheres, ChA activity continued to increase for a period of 3 months after hatching. Thereafter, enzymatic activity sharply declined and reached embryonic levels again by 3 years of age. ChA activity in the cerebellum increased for a period of up to 6 weeks after hatching, and in the diencephalon/midbrain region for a period of up to 3 months. If ChA activity is considered as signifying the presence of cholinergic neurons, then these data suggest differential rates of maturation for cholinergic neurons in various areas of the central nervous system. Although embryonic ages earlier than 14 days were not examined in this study, other studies in our laboratory (unpublished), and earlier studies by *Giacobini and Filogamo* (6), have detected ChA activity in the cerebral hemisphere of 6-day-old chick embryos.

The maturational profile of TH in embryonic chick brain differs from that of ChA. In both cerebral hemispheres and the diencephalon/midbrain region (fig. 2), TH activity was only barely detectable at 14 days of embryonic age. This activity increased sharply by 20 days of embryonic age, in the diencephalon/midbrain region; and by 1 day post-hatching in the cerebral hemispheres.

To further assess the maturational profiles of adrenergic neurons, we investigated the development of the uptake process for norepinephrine (NE) in the

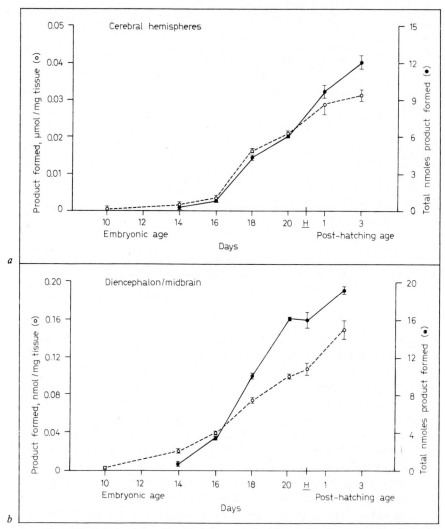

Fig. 2. Changes in TH activity during embryonic age (days) and post-hatching age (days) in two CNS areas: cerebral hemispheres (a) and diencephalon/midbrain (b). Points as in figure 1 (20).

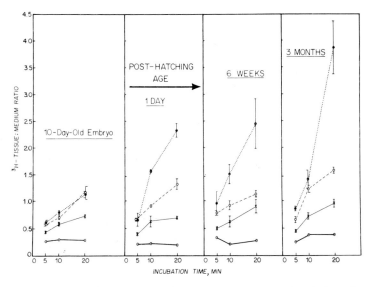

Fig. 3. Uptake of ^{3}H-NE, 1×10^{-7} M, into slices of cerebral hemispheres from 10-day-old chick embryos and from chicks at 1 day, 6 weeks, and 3 months post-hatching. The effects of cocaine (3×10^{-6} M) and reserpine (1×10^{-6} M) are illustrated: ⊙ = 0 °C control; ● = untreated; x = cocaine; ○ = reserpine. Points as in figure 1 (16).

chick brain. Prior to its storage within noradrenergic neurons, exogenous NE is taken up across the neuronal membrane and subsequently across the membrane of the granule within which it is ultimately stored. In the adult brain it is possible to demonstrate neuronal uptake of NE in slices (12), in isolated synaptosomes (3), and in homogenates (14). In our studies, uptake of NE was studied using slices of cerebral hemisphere tissue obtained from chick embryos or from chicks after hatching (fig. 3) (8, 16). A marked increase in the rate of accumulation of ^{3}H-NE (1×10^{-7} M) was observed in the cerebral hemisphere of 1-day-old chicks, compared to that seen in cerebral hemispheres of 10-day-old embryos. There was also a marked increase during the period from 1 day to 3 months after hatching. Incubation of the tissue slices in the presence of cocaine (3×10^{-5} M), an agent which inhibits neuronal uptake of NE, resulted in marked inhibition of ^{3}H-NE uptake at all ages. When reserpine (1×10^{-6} M), an agent which inhibits uptake of biogenic amines into storage granules, was added to the incubation medium, ^{3}H-NE uptake was inhibited only in tissue obtained from 20-day-old embryos (not shown in figure 3) (8) and from chicks after hatching. From these data, it appears that uptake processes in the cerebral hemispheres develop earlier than do the mechanisms for storage of NE. Since it also appears that the development of the uptake process for NE precedes that of tyrosine hydroxylase activity, the intraneuronal presence of NE may be important for the activation of this enzyme. The hypothesis that neurohumoral substances such as catecholamines, 5-hydroxytryptamine, and acetylcholine are involved in certain aspects of neural growth, including protein synthesis, has been supported by several investigators. This topic is addressed by *Lauder and Krebs* (pp. 171–180) and has been reviewed by *Vernadakis and Gibson* (18).

Neural Cell Culture Studies

We are currently investigating the regulation of the maturation of the neural enzymes ChA and TH using dissociated cell cultures of embryonic chick brain, as described by *Sensenbrenner et al.* (13). In this culture system the embryonic brain is sieved through nylon mesh (73 µm pore size) and the dispersed cells are plated in Falcon plastic flasks, at a density of 3 million cells/25 cm^2 flask. The cells are plated in 4 ml of Dulbecco's modified Eagle's medium fortified with 20% fetal calf serum. Within 24 h the dispersed cells form aggregates which attach to the surface of the flask. By day 5 in culture, neuroblasts and neurites can be observed. During the first 10 days, the cultures consist predominantly of neurons, with only a small number of glial cells present; after 15 days in culture, glial cells predominate.

ChA activity was measured in cultures of cerebral hemispheres obtained from 8-day-old chick embryos (fig. 4), using the method of *Fonnum* (5). The time course of the development of ChA activity was biphasic in nature, reaching a peak at day 10 in culture, declining between days 10 and 20, and increasing again from days 20 to 28 in culture. The early increase in ChA activity may reflect the differentiation of cholinergic neuroblasts in culture. The subsequent decline in enzymatic activity between days 10 and 20 may reflect a dilution of ChA activity by the proliferation of cells not containing ChA, such as adrenergic neurons, immature glial cells, and other nonneuronal cells. The later increase in enzyme activity may be in some manner related to the progressive maturation of glial cells during this time period. There is evidence that contact with glial cells enhances and maintains the differentiation of neurons in culture. For example, *Monard et al.* (9) have demonstrated the release of a macromolecular factor by glial cells in culture which can induce morphological differentiation of neuroblastoma cells, and *Murphy et al.* (10) have reported that cultured glial cells secrete a factor which is similar biologically and immunologically to nerve growth factor (NGF). In addition to possible glial influences, the increase in ChA activity seen in dissociated cultures may be related to the proliferation of other types of nonneuronal cells. *Young et al.* (21) have shown that primary cultures of chick fibroblasts secrete NGF. Finally, the possibilities that mature glial cells in this culture system contain ChA, or that undefined neuronal elements in the

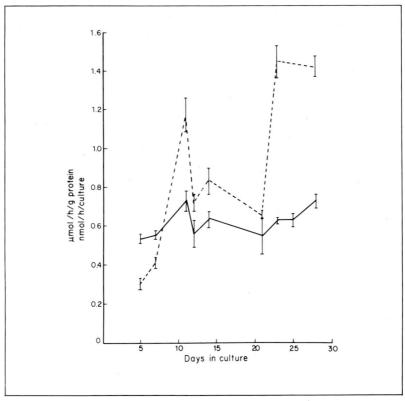

Fig. 4. Changes in ChA activity in 8-day-old chick embryo dissociated brain cells with days in culture. Points as in figure 1. Acetylcholine synthesized: ——— = per gram protein, ---- = per culture.

cultures are contributing to the elevation of ChA activity, cannot be discounted without further study.

The maturational profile of TH activity in culture differs from that of ChA activity. In both cerebral hemispheres and whole brain, there is a rapid increase in TH activity between 10 and 15 days in culture (fig. 5). These data, together with the data concerning the development of ChA activity, indicate that the capacity for enzymatic synthesis of acetylcholine appears earlier in culture than the capacity for catecholamine synthesis. This observation is in agreement with the *in vivo* studies described earlier. In the peripheral nervous system, presynaptic cholinergic neurons regulate the development of postsynaptic adrenergic neurons, as has been demonstrated by *Black et al.* (1, 2). *Otten and Thoenen* (11) have reported that acetylcholine or carbamylcholine can selectively induce TH and dopamine-β-hydroxylase in organ cultures of sympathetic ganglia.

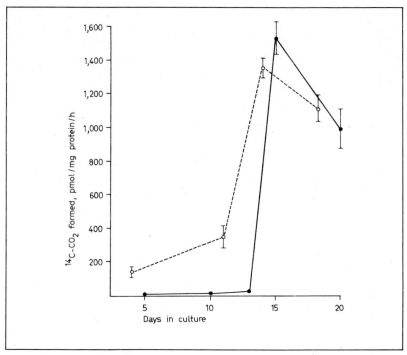

Fig. 5. Changes in TH activity in 8-day-old chick embryo dissociated brain cells with days in culture. Points as in figure 1. ——— = whole brain (– optic lobe), ----- = cerebral hemispheres.

Whether similar interactions exist between cholinergic and adrenergic neurons in dissociated brain cell cultures remains to be determined.

To further examine the maturation of adrenergic neurons in culture, the development of the uptake process for NE was studied. Dissociated cultures of whole brain (minus optic lobes) were incubated for 10 min with ^{3}H-NE (5×10^{-9} M). Cultures were prepared from 8-day-old embryos and uptake was determined following 5, 10, 15, and 20 days in culture. The accumulation of ^{3}H-NE, expressed as dpm per culture, increased between 5 and 10 days in culture and between 15 and 20 days in culture (not shown). When ^{3}H-NE accumulation was expressed as dpm/mg protein, the accumulation could be seen to decrease with age in culture, reflecting, perhaps, the proliferation of cells which are unable to accumulate NE. In an attempt to distinguish between neuronal and extraneuronal accumulation of NE, cultures of whole brain from 8-day-old chick embryos were preincubated for 10 min with either desmethylimipramine (DMI), an agent which inhibits neuronal uptake; or metanephrine, an agent which inhibits extraneuronal uptake (7). The accumulation of ^{3}H-NE was

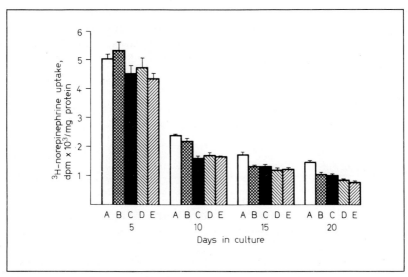

Fig. 6. Uptake of ^{3}H-NE, 5×10^{-9} M, 10 min incubation, in chick embryo dissociated brain cells with days in culture. The effects of DMI are illustrated: A = control; B = pre-incubated for 10 min with DMI, 10^{-7} M; C = DMI, 10^{-6} M; D = DMI, $10^{-5}$$M$; E = DMI, 10^{-4} M. Points with bracketed lines represent means and SEM.

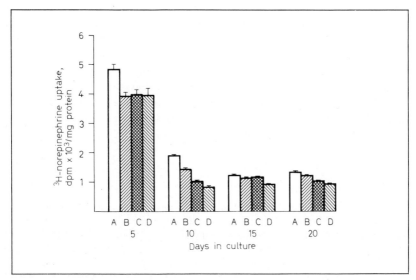

Fig. 7. As in figure 6, except that cultures were pre-incubated for 10 min with metanephrine, prior to uptake of ^{3}H-NE: A = control; B = metanephrine, 10^{-6} M; C = metanephrine, 10^{-5} M; D = metanephrine, 10^{-4} M.

inhibited by DMI as early as 5 days in culture, although the inhibition was significant only at high concentrations (fig. 6). In contrast, the inhibition of ^{3}H-NE accumulation by metanephrine was very pronounced in 5-day-old cultures, and remained significant at 10 and 20 days in culture (fig. 7). Inhibition by metanephrine during early stages of growth in culture (up to 10 days), may represent primarily inhibition of uptake of NE by neuroblasts, immature neurons, and presumptive glial cells; while at later stages of growth in culture (20 days) the inhibition may represent inhibition of uptake of NE by glial cells.

Conclusions

The maturational profiles of two neural enzymes, ChA and TH, and the development of the uptake process for NE were compared between embryonic chick brain tissue and dissociated cell cultures of embryonic chick brain. The maturational profiles of the two enzymes in these systems are similar, in that ChA activity appears earlier than TH activity, both during embryonic development and during the growth of cultured embryonic brain cells. Moreover, in both systems the uptake process for NE is operative at an earlier stage in development than is the capacity for synthesis of NE. These findings would tend to support the suggestion that the neuronal uptake of NE may be important for the activation of the enzyme and perhaps for the maturation of the adrenergic neuron. The fact that the appearance of the cholinergic enzyme, ChA, precedes that of TH may also suggest that cholinergic mechanisms may be involved in the maturation of the adrenergic neuron, as is the case in the peripheral sympathetic nervous system.

The findings reported here demonstrate the usefulness of neural tissue culture as a model for the study of maturational aspects of neurotransmission. With such systems, factors regulating neurotransmission may be more clearly defined and manipulated, thus facilitating elucidation of cellular and molecular mechanisms involved in the ontogeny of neurotransmission.

Summary

The maturational profiles of two neural enzymes, choline acetyltransferase (ChA) and tyrosine hydroxylase (TH), and the development of the uptake process for norepinephrine (NE) were compared between embryonic chick brain tissue and dissociated cell cultures of embryonic chick brain. The developmental profiles of the two enzymes in these systems are similar, in that ChA activity appears earlier than TH activity, both during embryonic development *in vivo* and during the growth of cultured embryonic brain cells. Moreover, in both systems the uptake process for NE is operative at an earlier stage in development than is the capacity for synthesis of NE. The findings reported here demonstrate the usefulness of neural tissue culture as a model for the study of maturational aspects of neurotransmission.

References

1. *Black, I.B.; Bloom, E.M., and Hamill, R.W.:* Central regulation of sympathetic neuron development. Proc. natn. Acad. Sci. USA *73:* 3575–3578 (1976).
2. *Black, I.B.; Hendry, I.A., and Iversen, L.L.:* Transsynaptic regulation of growth and development of adrenergic neurons in a mouse sympathetic ganglion. Brain Res. *34:* 229–240 (1971).
3. *Davis, J.M.; Goodwin, F.K.; Bunney, W.E.; Murphy, D.L., and Colburn, R.W.:* Effects of ions in uptake of norepinephrine by synaptosomes. Pharmacologist *9:* 184 (1967).
4. *Fellman, J.H.:* A chemical method for the determination of acetylcholine. Its application in a study of presynaptic release and choline acetyltransferase assay. J. Neurochem. *16:* 135–143 (1969).
5. *Fonnum, F.:* A rapid radiochemical method for the determination of choline acetyltransferase. J. Neurochem. *24:* 407–409 (1975).
6. *Giacobini, G. and Filogamo, G.:* Changes in the enzymes for the metabolism of acetylcholine during development of the central nervous system; in *Genazzani and Herken* Central nervous system studies on metabolic regulation and function, pp. 153–157 (Springer, Berlin 1973).
7. *Iversen, L.L.:* Role of transmitter uptake mechanisms in synaptic neurotransmission. Br. J. Pharmacol. *41:* 571–591 (1971).
8. *Kellog, C.; Vernadakis, A., and Rutledge, C.O.:* Uptake and metabolism of ^{3}H-norepinephrine in the cerebral hemisphere of chick embryos. J. Neurochem. *18:* 1931–1938 (1971).
9. *Monard, D.; Solomon, F.; Rentsch, M., and Gysin, R.:* Glia-induced morphological differentiation in neuroblastoma cells. Proc. natn. Acad. Sci. USA *70:* 1894–1897 (1973).
10. *Murphy, R.A.; Oger, J.; Saide, J.D.; Blanchard, M.H.; Aranson, G.W.; Hogan, C.; Pantazis, N.J., and Young, M.:* Secretion of nerve growth factor by central nervous system glioma cells in culture. J. Cell Biol. *72:* 769–773 (1977).
11. *Otten, U. and Thoenen, H.:* Mechanisms of tyrosine hydroxylase and dopamine-β-hydroxylase induction in organ cultures of rat sympathetic ganglia by potassium depolarization and cholinomimetics. Archs Pharmacol. *292:* 153–159 (1976).
12. *Rutledge, C.O.:* The mechanisms by which amphetamine inhibits oxidative deamination of norepinephrine in brain. J. Pharmac. exp. Ther. *171:* 188–195 (1970).
13. *Sensenbrenner, M.; Booher, J., and Mandel, P.:* Cultivation and growth of dissociated neurons from chick embryo cerebral cortex in the presence of different substrates. Z. Zellforsch. mikrosk. Anat. *117:* 559–569 (1971).
14. *Snyder, S.H. and Coyle, J.T.:* Regional differences in ^{3}H-norepinephrine and ^{3}H-dopamine uptake into rat brain homogenates. J. Pharmac. exp. Ther. *165:* 78–86 (1969).
15. *Vernadakis, A.:* Comparative studies of neurotransmitter substances in the maturing and aging nervous system of the chicken; in *Ford* Neurobiological aspects of maturation and aging, pp. 231–243 (Elsevier, Amsterdam 1973).
16. *Vernadakis, A.:* Uptake of ^{3}H-norepinephrine in the cerebral hemispheres and cerebellum of the chicken throughout the life span. Mechanisms Aging Develop. *2:* 371–379 (1973).
17. *Vernadakis, A. and Culver, B.:* Neural tissue culture. A biochemical tool; in *Kumar* Biochemistry of the brain (Pergamon Press, London, in press 1978).
18. *Vernadakis, A. and Gibson, D.A.:* Role of neurotransmitter substances in neural growth; in *Dancis and Hwang* Perinatal pharmacology: problems and priorities, pp. 65–77 (Raven Press, New York 1974).

19 Waymire, J.C.; Bjur, R., and Weiner, N.: Assay of tyrosine hydroxylase by coupled decarboxylation of dopa formed from 1-^{14}C-L-tyrosine. Analyt. Biochem. *43:* 588–600 (1971).
20 Waymire, J.C.; Vernadakis, A., and Weiner, N.: Studies on the development of tyrosine hydroxylase, monoamine oxidase and aromatic-L-amino acid decarboxylase in several regions of the chick brain; in *Vernadakis and Weiner* Drugs and the developing brain, pp. 149–170 (Plenum Publishing, New York 1974).
21 Young, M.; Oger, J.; Blanchard, M.H.; Asdourian, H.A., and Aranson, B.G.W.: Secretion of a nerve growth factor by primary chick fibroblast cultures. Science *187:* 361–362 (1975).

A. Vernadakis, PhD, Departments of Psychiatry and Pharmacology, University of Colorado, *Denver, CO 80262* (USA)

Serotonin and Early Neurogenesis[1]

Jean M. Lauder and Helmut Krebs

Department of Biobehavioral Sciences, University of Connecticut, Storrs, Conn. and A. Ribicoff Research Center, Norwich Hospital, Norwich, Conn.

Introduction

The concept that the monoamines and other neurotransmitter substances might play a role in early embryogenesis prior to the onset of neurotransmission was first suggested by *Buznikov et al.* (13) in the 1960s. Since then, a number of developmental neurobiologists have proposed possible ontogenetic functions for these neurohumoral agents (7, 8, 24, 29, 36, 41, 42).

Recently, we raised the possibility that the monoamines norepinephrine (NE), dopamine (DA) and serotonin (5-HT) might be important in regulating the onset of differentiation (cessation of proliferation) of their ultimate target cells in the neural tube (25). This idea was drawn from the facts that the monoamine neurons begin their differentiation very early in gestation and appear to be able to synthesize transmitter and elaborate processes soon thereafter (25, 30), even though they are not innervated until just prior to birth (26). Moreover, the cells of the locus ceruleus start to differentiate several days prior to their target cells (fig. 1a). At the time when monoamine cell differentiation is beginning (days 10–11 of gestation) the rat neural tube is just closing (17, 43). Since the first postmitotic neuroblasts are thought to form just after neural tube closure (23), these monoamine cells must be among the first group of cells to begin differentiation. If such early forming neurons were to be influential in controlling the timing of last cell division of their eventual target cells, either by directly triggering germinal cells to cease dividing, or by sensitizing them to some other differentiating agent, while 'marking' them for future recognition during synaptogenesis, these cells would be direct participants in the molding of their own circuitry in the developing brain.

[1] Part of this work was performed in the Laboratory of Neuropharmacology, NIMH, St. Elizabeth's Hospital, Washington, D.C. These studies were also partially supported by NIH grant number NS-13481-01 to J.M.L.

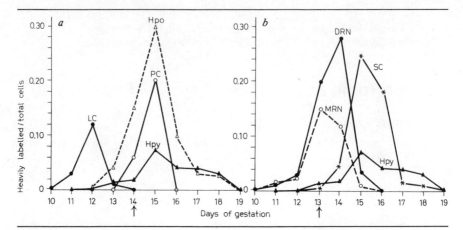

Fig. 1. Time course of onset of neuronal differentiation (cessation of germinal cell proliferation) in locus ceruleus (LC) (a) and dorsal (DRN) and medial (MRN) raphe nuclei (b) and some of their known target cells, hippocampal pyramidal (Hpy) and polymorph cells (Hpo), and cells of the superior colliculus, stratum griseum superficiale (SC). Arrows: day when fluorescence is first detectable in these monoamine-containing cell groups. *a* From *Lauder and Bloom* (25). *b* From *Lauder and Krebs* (28).

To test the hypothesis that the monoamines might be important for the onset of differentiation of their target cells, we chose the 5-HT neurons as a model, since a pharmacological tool was available (*p*-chlorophenylalanine, pCPA) which could deplete 5-HT in axons and terminals without destroying them (4, 15, 22), a feature not present with the catecholamine-depleting drug, 6-hydroxydopamine. 5-HT is also of particular interest since it has been implicated in various aspects of early embryogenesis such as cleavage and gastrulation (7, 12, 13, 16) which involve cell motility and contractile elements of the cell (7, 13, 20), features also important in the development of the neural tube. In the rat, 5-HT is demonstrable biochemically 1 week prior to birth (7, 14), and 5-HT fluorescence can be detected in developing 5-HT neurons and their processes early in gestation (days 12–13), when a large proportion of prospective 5-HT

Fig. 2. Effects of pCPA and 'stress' on the onset of neuronal differentiation in brain regions known to be innervated by 5-HT terminals (5-HTT) (3, 4, 18). CAT = Catecholamine terminals (18); AT = adrenaline terminals (22); 0 = not identified terminals; S = scattered, few terminals; + 1 to + 5 = relative numbers of terminals. ∗——∗ = Uninjected 'normal' controls (UC); ●——● = injected 'stressed' controls (IC); □·········□ = pCPA injected (P); ★ = IC or P significantly different from UC at p <0.01; · = IC significantly different from P at p <0.01 (Duncan's multiple F-test). Brain region abbreviations: SC = Superior

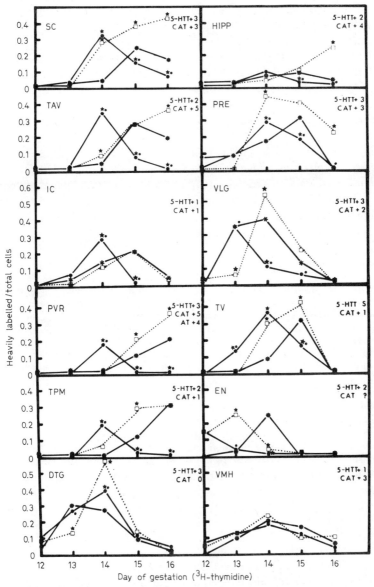

colliculus, stratum griseum superficiale; HIPP ventral hippocampal pyramidal cells; TAV = thalamic nucleus anterior ventralis; PRE = pretectal nucleus; IC = inferior colliculus; VLG = ventral lateral geniculate; PVR = thalamic nucleus periventricularis rotundocellularis; TV = thalamic nucleus ventralis; TPM = thalamic nucleus posteromedianus; EN = entopeduncular nucleus; DTG = dorsal tegmental nucleus (Gudden); VMH = hypothalamic ventromedial nucleus. From *Lauder and Krebs* (28).

cells are still dividing. It is not clear whether 5-HT is synthesized in such cells prior to the cessation of cell division (25, 30) and may actually be an internal differentiating agent in cells which contain it (25). Of particular interest for our hypothesis is the fact that 5-HT fluorescence is detectable at the very beginning of differentiation of 5-HT target cells, which follow the 5-HT containing raphe neurons by several days (fig. 1b), as noted with the locus ceruleus and its target cells.

In an attempt to demonstrate that 5-HT depletion could lead to specific retardation of 5-HT target cell differentiation, we employed the 'morphochemical' tool of long survival ^{3}H-thymidine autoradiography for dating time of origin (last cell division) of neurons (5, 38) to assess the effects of pCPA treatment of the pregnant rat on fetal brain development in those regions known to contain 5-HT terminals (5-HT target cells), as compared to areas where no 5-HT terminals have been identified.

Methods and Materials

As described in detail elsewhere (27), we injected pCPA-methyl-ester-HCl into pregnant rats intraperitoneally (P) beginning on day 8 of gestation (day 1 = day of insemination), shortly after the blastocyst has completed implantation and come under the influence of the maternal circulation, although the placenta has not yet developed (33, 43). Control mothers were either injected with vehicle (Ringer's solution, injected controls, IC), or uninjected (UC). All treatments were continued until the day of ^{3}H-thymidine injection on day 12, 13, 14, 15 or 16 of gestation. Pups were allowed to survive for 30 days following birth, at which time they were perfused under anesthesia, their brains removed and prepared for autoradiography.

Counts of heavily labelled cells (neurons which began to differentiate on the day of isotope injection, i.e., became postmitotic) were made in 45 brain regions either known to receive 5-HT innervation or thought to contain no 5-HT terminals. Heavily labelled/total cells was used as a measure of the proportion of the neuronal population beginning to differentiate on a particular day of gestation. By this method, a time course for cell differentiation was obtained which could be used to assess the effects of the treatments (fig. 2, 3).

Results

The effects of pCPA treatment of the pregnant rat on the beginning of neuronal differentiation in the fetal brain are illustrated in figures 2 and 3.

Consistent with our hypothesis that 5-HT is important for the timing of differentiation of 5-HT target cells, figure 2 illustrates that the effects of pCPA are greatest in *5-HT terminal regions,* manifested as delayed (initially suppressed), late and/or prolonged peaks when compared to IC. Compared to UC, however, effects of pCPA are more variable, ranging from delayed, late and/or

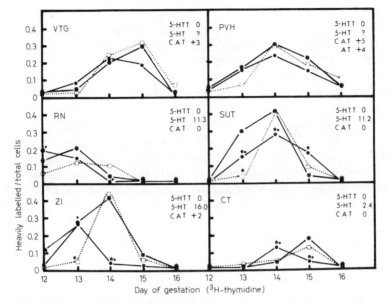

Fig. 3. Effects of pCPA and 'stress' on the onset of neuronal differentiation in brain regions where no 5-HT terminals (5-HTT) have been found (3, 4, 18). CAT = Catecholamine terminals (19); AT = adrenaline terminals (21); 5-HT = biochemically measured 5-HT in ng/mg protein (31, 34, 35); 0 = not identified terminals; + 2 to + 5 = relative numbers of terminals. *———* = Uninjected 'normal' controls, UC; •———• = injected 'stressed' controls, IC; ▫·········▫ = pCPA injected, P; ★ = IC or P significantly different from UC at p <0.01; · = IC significantly different from P at p <0.01 (Duncan's multiple F-test). Brain region abbreviations: VTG = Ventral tegmental nucleus (Gudden); PVH = hypothalamic paraventricular nucleus; RN = red nucleus; SUT = subthalamic nucleus; ZI = zona incerta; CT = trapezoid body. From *Lauder and Krebs* (28).

prolonged peaks to early peaks or peaks with early beginnings. In a number of cases, no differences in the peaks of P and UC are seen, except that the P peak is frequently elevated.

When the three treatment groups are compared it is obvious that the offspring of IC mothers exhibit consistently earlier onset of differentiation in *5-HT terminal regions* (fig. 2) when compared to UC or pCPA injected animals. This pattern is also seen in regions with *no 5-HT terminals,* but relatively high contents of 5-HT, measured biochemically (31, 34, 35) or in areas which contain catecholamine (CA) terminals (fig. 3). The only exception to this rule is the trapezoid body (CT: fig. 3), which contains neither CA terminals, nor high levels of 5-HT. The early onset of differentiation in IC is attributed to the chronic 'stress' effects of the daily vehicle injections, as discussed in detail below.

In *non 5-HT terminal regions,* several instances are seen where there are no significant differences between the three treatment groups (VTG, PVH; fig. 3), an expected finding if the differentiation of 5-HT target cells is selectively affected by changes in 5-HT levels. It should be noted, however, that some similar cases are also found in areas containing 5-HT terminals (VMH, MRN; fig. 2), perhaps signifiying that 5-HT is not important in the development of these regions.

Where effects are seen in nonterminal areas, they are of a far lesser magnitude than in terminal regions, with all three peaks frequently overlapping. In those instances where the pCPA peak is delayed (SUT, ZI; fig. 3), these regions contain relatively high levels of 5-HT, indicating that terminals may actually be present, but were not detected in fluorescence histochemical studies.

Discussion

Our studies have demonstrated that treatment of the pregnant rat with the 5-HT depleting drug pCPA results in retarded onset of neuronal differentiation in those fetal brain regions which, in the adult, will contain 5-HT nerve terminals or have high contents of 5-HT, indicating that these cells represent future 5-HT target neurons. This finding provides support for our hypothesis that 5-HT is an important regulator of the cessation of germinal cell proliferation in the neural tube, specifically those cells with which 5-HT terminals will eventually form synaptic contacts. Thus, 5-HT may influence the beginning of differentiation of specific neurons, while at the same time 'marking' them for future recognition during synaptogenesis.

Scanning electron microscopic studies of the developing cerebral vesicle (37) have shown that at day 13, when cell differentiation is peaking in the 5-HT containing raphe neurons, and just beginning in many of their target cells (fig. 1, 2), the neural tube is becoming a complex, heterogeneous population of postmitotic neuroblasts and proliferating neuroepithelial cells, where a myriad of opportunities for cell-cell interactions exist. This is in marked contrast to day 12 when the neural tube essentially consists of a proliferating neuroepithelium. In such a system it is not difficult to imagine that local interactions between 5-HT neuroblasts and undifferentiated germinal cells could occur, providing the first step in the construction of the 5-HT circuitry.

Although we have not yet examined whether 5-HT is depleted in the fetal brain, particularly in developing 5-HT axons, as a result of pCPA treatment of the mother, the specificity of the retardation effects of pCPA leads us to believe that this may actually result from reduced levels of 5-HT at a time when it is required as a 'differentiation signal' in the neural tube.

The early differentiation of 5-HT target neurons observed in offspring of IC mothers is particularly interesting in view of the rapidly emerging literature linking *glucocorticoids* to the regulation of monoamine synthesis, particularly in young animals, and the ability of stress to mimic these effects (6, 39). If stress-induced elevations in maternal glucocorticoids influence the fetus, resulting in increased levels of 5-HT, this could produce effects essentially opposite to those of pCPA (5-HT depletion), as observed in our studies.

Glucocorticoid-mediated stress effects would appear to be a viable possibility given the facts that this hormone readily passes the placenta (44), as well as being secreted by it (9); chronic stress appears to produce elevated plasma levels of plasma glucocorticoids in the pregnant mouse (10); and high doses of glucocorticoids can cause teratogenic effects in fetal mice when administered during the same 'critical period' covered in our studies (32). Since the fetal adrenal undergoes organogenesis on days 12–16 in the fetal rat and is capable of producing glucocorticoids by day 14, several days prior to coming under the regulatory influence of the pituitary (19), it is possible that this early production of hormone is important in regulating 5-HT levels in the fetal brain, and thus the rate of differentiation of 5-HT target neurons.

Our results may also be relevant to reported effects of prenatal handling or sham injections on the postnatal susceptibility to stress, where offspring of handled or injected mothers were more emotional (1), exhibited accelerated development of circadian rhythmicity of adrenal function (2), and were more susceptible to audiogenic seizures (11), which have been linked to glucocorticoids and altered 5-HT metabolism in recent studies (39, 40).

The effects of pCPA and 'stress' demonstrated in our studies should emphasize the possible adverse effects of monoamine-interactive drugs, such as antidepressants and tranquillizers, and hormones such as glucocorticoids administered to the pregnant woman on early brain development. Environmental stress, especially in chronic form, would appear to be an equally dangerous influence on the fetal brain.

While these drugs, hormones, and stressors may not produce gross teratogenic effects, they may offset the timing of development of specific brain regions, leaving others relatively unaffected, thus changing the wiring of the developing brain circuitry. Such changes may be severe enough to be noticed as mental retardation or hyperactivity, or be exhibited as an altered level of intelligence or emotionality in an individual who would otherwise have fallen within the normal range.

Summary

Experimental evidence is presented to support the hypothesis that the *monoamines*, particularly *serotonin*, act as *humoral signals* in the early neural tube for the onset of *differentiation* of those cells which will eventually receive monoaminergic innervation *(monoamine target cells)*. By implication, at the same time the monoamines act as 'differentiation signals', they may also 'mark' these cells for future recognition during synaptogenesis.

Hormones, such as the *glucocorticoids*, may regulate the timing of this process by controlling the rate or amount of monoamine synthesis in the fetal brain.

Stress or monoamine-interactive *drugs*, administered to the *pregnant female*, could affect fetal brain development by altering the onset of monoamine target cell differentiation.

References

1. *Ader, R. and Conklin, P.M.:* Handling of pregnant rats. Effects on emotionality of their offspring. Science *142:* 411–412 (1963).
2. *Ader, R. and Deitchman, R.:* Effects of prenatal maternal handling on the maturation of rhythmic processes. J. comp. physiol. Psychol. *71:* 492–496 (1970).
3. *Aghajanian, G.K.; Haigler, H.L., and Bennett, J.L.:* Amine receptors in CNS. III. 5-Hydroxytryptamine in brain; in Iversen, Iversen and Snyder Handbook of psychopharmacology, vol. 6 (Plenum Press, New York 1974).
4. *Aghajanian, G.K.; Kuhar, M.J., and Roth, R.H.:* Serotonin-containing neuronal perikarya and terminals: differential effects of p-chlorophenylalanine. Brain Res. *54:* 85–101 (1973).
5. *Altman, J.:* DNA metabolism and cell proliferation; in Lajtha Handbook of neurochemistry, vol. 2, pp. 137–182 (Plenum Press, New York 1969).
6. *Azmitia, E.C. and McEwen, B.S.:* Adrenalcortical influence on rat brain tryptophan hydroxylase activity, Brain Res. *78:* 291–302 (1974).
7. *Baker, P.C. and Quay, W.B.:* 5-Hydroxytryptamine metabolism in early embryogenesis and the development of brain and retinal tissues. A review. Brain Res. *12:* 273–295 (1969).
8. *Balázs, R.:* Influence of metabolic factors on brain development. Br. med. Bull. *30:* 126–134 (1974).
9. *Barlow, S.M.; Morrison, P., and Sullivan, F.M.:* Plasma corticosterone levels during pregnancy in the mouse: the relative contributions of the adrenal glands and foetal-placental units. J. Endocr. *60:* 473–483 (1974).
10. *Barlow, S.M.; Morrison, P., and Sullivan, F.M.:* Effects of acute and chronic stress on plasma corticosterone levels in pregnant and non-pregnant mouse. J. Endocr. *66:* 93–99 (1975).
11. *Beck, S.L. and Gavin, D.L.:* Susceptibility of mice to audiogenic seizures is increased by handling their dams during gestation. Science *193:* 427–428 (1976).
12. *Buznikov, G.A. and Berdysheva, L.V.:* Changes in functional activity of the neurohormones in embryos of *Paracentrotus lividus* during the mitotic cycle. Dokl. Akad. Nauk. SSSR Otd. Biol. *167:* 486–488 (1966).
13. *Buznikov, G.A.; Chudakova, I.V., and Zvebina, N.D.:* The role of neurohumors in early embryogenesis. I. Serotonin content of developing sea urchin and loach. J. Embryol. exp. Morph. *12:* 563–573 (1964).
14. *Coyle, J.T. and Henry, D.:* Catecholamines in fetal and newborn rat brain. J. Neurochem. *21:* 61–67 (1973).
15. *Deguchi, T.; Sinha, A.K., and Barchas, J.D.:* Biosynthesis of serotonin in raphe nuclei of rat brain: effect of p-chlorophenylalanine. J. Neurochem. *20:* 1329–1336 (1973).
16. *Emanuelsson, H.:* Localization of serotonin in cleavage embryos of *Ophryotrocha labronica* La Greca and Bacci. Rouxs Arch. EntwMech. Org. *175:* 253–271 (1974).

17 *Freeman, B.G.:* Surface modifications of neural epithelial cells during formation of the neural tube in the rat embryo. J. Embryol. exp. Morph. *28:* 437–448 (1972).
18 *Fuxe, K.:* Evidence for the existence of monoamine neurons in the central nervous system. IV. Distribution of monoamine nerve terminals in the central nervous system. Acta physiol. scand. *64:* suppl. 247, pp. 39–85 (1965).
19 *Greengard, O.:* The developmental formation of enzymes in rat liver; in *Litwack* Biological actions of hormones, vol. 1, pp. 53–87 (Academic Press, New York 1970).
20 *Gustafson, T. and Toneby, M.:* On the role of serotonin and acetylcholine in sea urchin morphogenesis. Expl Cell Res. *62:* 102–117 (1970).
21 *Hökfelt, T.; Fuxe, K.; Goldstein, M., and Johansson, O.:* Evidence for adrenaline neurons in the rat brain. Acta physiol. scand. *89:* 286–288 (1973).
22 *Koe, B.K. and Weissman, A.:* The pharmacology of para-chlorophenylalanine, a selective depletor of serotonin stores. Adv. Pharmacol. *6B:* 29–47 (1968).
23 *Langman, J.; Guerrant, R.L., and Freeman, B.G.:* Behavior of neuroepithelial cells during closure of the neural tube. J. comp. Neurol. *127:* 399–412 (1966).
24 *Lanier, L.P.; Dunn, A.J., and Van Hartesveldt, C.:* Development of neurotransmitters and their function in brain; in *Ehrenpreis and Kopin* Reviews of neuroscience, vol. 2, pp. 195–256 (Raven Press, New York 1976).
25 *Lauder, J.M. and Bloom, F.E.:* Ontogeny of monoamine neurons in the locus coeruleus, raphe nuclei and substantia nigra of the rat. I. Cell differentiation. J. comp. Neurol. *155:* 469–481 (1974).
26 *Lauder, J.M. and Bloom, F.E.:* Ontogeny of monoamine neurons in the locus coeruleus, raphe nuclei and substantia nigra of the rat. II. Synaptogenesis. J. comp. Neurol. *163:* 251–264 (1975).
27 *Lauder, J.M. and Krebs, H.:* Effects of p-chlorophenylalanine on time of neuronal origin during embryogenesis in the rat. Brain Res. *107:* 638–644 (1976).
28 *Lauder, J.M. and Krebs, H.:* Serotonin as a differentiation signal in early neurogenesis. Dev. Neurosci. (in press, 1978).
29 *McMahon, D.:* Chemical messengers in development: a hypothesis. Science *185:* 1012–1021 (1974).
30 *Olson, L. and Seiger, Å:* Early prenatal ontogeny of central monoamine neurons in the rat: fluorescence histochemical observations. Z. Anat. EntwGesch. *137:* 301–316 (1972).
31 *Palkovits, M.; Brownstein, M., and Saavedra, J.M.:* Serotonin content of the brain stem nuclei in the rat. Brain Res. *80:* 237–249 (1974).
32 *Pinsky, L. and Digeorge, A.M.:* Cleft palate in the mouse: a teratogenic index of glucocorticoid potency. Science *147:* 402–403 (1965).
33 *Rugh, R.:* The mouse: its reproduction and development, pp. 85–107 (Burgess, Minneapolis 1968).
34 *Saavedra, J.M.; Palkovits, M.; Brownstein, M.J., and Axelrod, J.:* Serotonin distribution in the nuclei of the rat hypothalamus and preoptic region. Brain Res. *77:* 157–165 (1974).
35 *Saavedra, J.M.; Brownstein, M., and Palkovits, M.:* Serotonin distribution in the limbic system of the rat. Brain Res. *79:* 437–441 (1974).
36 *Seiger, Å and Olson, L.:* Late prenatal ontogeny of central monoamine neurons in the rat: fluorescence histochemical observations. Z. Anat. EntwGesch. *140:* 281–318 (1973).
37 *Seymour, R.M. and Berry, M.:* Scanning and transmission electron microscope studies of interkinetic nuclear migration in the cerebral vesicles of the rat. J. comp. Neurol. *160:* 105–126 (1975).

38 *Sidman, R.L.:* Autoradiographic methods and principles for the study of the nervous system with thymidine-^{3}H; in *Nauta and Ebbesson* Contemporary research methods in neuroanatomy, pp. 252–274 (Springer, New York 1970).
39 *Sze, P.Y.:* Glucocorticoid regulation of the serotonergic system of the brain; in *Costa, Giacobini and Paoletti* First and second messengers: new vistas. Advances in biochemical psychopharmacology, vol. 15, pp. 251–265 (Raven Press, New York 1976).
40 *Sze, P.Y. and Maxson, S.C.:* Involvement of corticosteroids in acoustic induction of audiogenic seizure susceptibility in mice. Psychopharmacologia *45:* 79–82 (1975).
41 *Toneby, M.:* Functional aspects of 5-hydroxytryptamine and dopamine in early embryogenesis of *Echinoidea* and *Asteroidea* (Wenner-Grenn Institute, Stockholm 1977).
42 *Vernadakis, A. and Gibson, D.A.:* Role of neurotransmitter substances in neural growth; in *Dancis and Hwang* Perinatal pharmacology: problems and priorities, pp. 65–76 (Raven Press, New York 1974).
43 *Witschi, E.:* in Growth. Biological handbooks, pp. 304–314 (Fed. Am. Soc. exp. Biol., Baltimore 1962).
44 *Zarrow, M.X.; Haltmeyer, G.C.; Denenberg, V.H., and Thatcher, J.:* Response of the infantile rat to stress. Endocrinology *79:* 631–634 (1966).

J.M. Lauder, PhD, Department of Biobehavioral Sciences, University of Connecticut, *Storrs, CT 06268* (USA)

Adaptive Changes Induced by Environmental and Hormonal Factors on the Development of Brain Neurotransmitter Systems

Paola S. Timiras and Andrea Vaccari

Department of Physiology-Anatomy, University of California, Berkeley, Calif., and Department of Pharmacology, University of Genova, Genova

Introduction

Living organisms can adjust their internal environment to meet external challenges through neuroendocrine adaptations. Thus, survival depends on both nervous (e.g., modulation of central and peripheral neurotransmitter systems) and endocrine (e.g., stimulation of the hypothalamic-hypophyseal-adrenocortical axis) responses and their integration of all body functions. Considering monoaminergic neurotransmitters and adrenocortical hormones central to these integrative regulations, the following sequence of events can be postulated: external stimuli increase adrenocortical secretion through stimulation of the hypothalamic-hypophyseal-adrenocortical axis; the ensuing increased hormonal levels influence monoaminergic neurotransmission by acting on the metabolic enzymes for monoamines; reciprocally, stimulation or inhibition of brain monoaminergic pathways and changes in hormonal levels regulate the release of hypothalamic, pituitary and adrenocortical hormones — the latter, in turn, reach the target tissues where they act as 'gene triggers' to regulate the function of effector and sensory cells, and, in some cases, to cause target tissues to undergo developmental changes as well. Each of the control systems, hypothalamus, pituitary, and peripheral endocrines, as well as the higher cerebral (cortical) centers, is regulated by a series of positive and negative feedbacks, which, for the maintenance of homeostasis, must achieve harmonious integration under steady-state and stress conditions.

The nature and efficiency of these control systems is well known in the adult but much less so during development. Experimental and clinical data suggest that the capacity to adapt is less in the young than in the adult and, indeed, may be so inadequate as to account for a perinatal 'stress-nonresponsive' period, the duration of which varies with the animal species. In the rat, which is very immature at birth, this period would coincide with the first postnatal week and may be ascribable to immaturity of the brain centers — particularly those ontogenetically less mature — rather than to immaturity of the pituitary and/or target endocrines. Among the maturational events which seem to play a key role in conferring to the animal the capability to adapt, are those concerned with synaptogenesis and maturation of neurotransmitter pathways. Thus, in the rat, the enzymes involved in brain neurotransmission are significantly less active at birth and during early development than at later ages and, for some of them, such as monoamine oxidase (MAO), activity continues to increase even after sexual maturity has been achieved (20). Age-related differences in enzyme activity and in the levels of neurotransmitters directly reflect synaptogenesis and proliferation of nerve endings during development. Conceivably any factor affecting the biochemical maturation of the brain will also affect the maturation of neuroendocrine integrative regulations and impair adaptation.

Here we will consider the following three questions: What is the response to external and internal stimuli of the immature as compared to the adult brain? What is the importance of the duration and severity of the stimulation on these responses? What are the mechanisms by which neuroendocrine adjustments take place during development and what is their significance for the efficiency of integrative functions and adaptive competence? To answer these questions, newborn and adult rats were exposed to a variety of unfavorable external environments (high altitude hypoxia, increased noise-light-motion, radiation) and changed internal conditions (hypothyroidism, sex differences) for short or prolonged periods of time. Under these conditions, the development of a number of neurotransmitter systems, particularly cholinergic and monoaminergic systems, were measured in discrete brain areas and, in some cases, in the adrenal medulla as well, from birth until adulthood.

High Altitude Hypoxia

Although several factors may intervene at natural high altitude and alter brain development, hypoxia appears to be the most significant in view of the oxygen dependence of brain metabolism (11, 14). That prolonged and/or acute exposure to hypoxia induces neurologic and behavioral alterations is supported by experimental and clinical data (11, 12, 14–16, 26). Indeed, exposure to anoxia or severe hypoxia is a well-known noxious event for human fetuses and neonates, which may lead to temporary or irreversible neurologic damage and behavioral disabilities (12, 15, 26). Acute or subacute exposure of developing and adult animals to severe hypoxia decreases the activity of monoaminergic synthesizing enzymes (primarily in whole brain) without, however, significantly altering monoamine levels (17).

Our experiments at the White Mountain Research Station of the University of California (altitude: 3,800 m; oxygen and barometric pressure: approximately 13% and $^1/_2$ of sea level) were designed to study whether exposure to a moderate hypoxic environment, prolonged throughout pre- and postnatal development, would induce adaptive responses in monoaminergic systems in the rat. Rats conceived, born and maintained before and after birth at high altitude displayed a high mortality rate, behavioral and neurologic symptoms and endocrine alterations (e.g., increased corticosterone levels (16). Monoaminergic pathways in the brain (1, 17) and the adrenals (19) were altered in their development depending on the age of the animal studied and the brain region and enzyme considered. Enzyme activity was generally less sensitive to hypoxia during the first postnatal week than at later ages; some areas, still immature at birth, such as the cerebral cortex and cerebellum, were significantly affected early in development (e.g., during the first 12 days of age, tyrosine hydroxylase (TH) activity was higher at high altitude than at sea level). On the other hand, the activities of tryptophan hydroxylase, TH, L-3,4-dihydroxyphenylalanine (DOPA) and 5-hydroxytryptophan (5-HTP) decarboxylases, MAO and catechol-O-methyltransferase (COMT) were all decreased by hypoxia in the hypothalamus of adult animals. A similar dual response of increased enzymatic activity at an early age followed by decreased activity at later ages in hypoxic as compared to normoxic rats was observed also in specific cholinergic enzymes (16). Whereas decarboxylase activity, the nonoxidative step in catecholamine synthesis, was apparently unaffected by acute hypoxia (2), our studies show that chronic hypoxia alters both oxygen-dependent (i.e., hydroxylases) and oxygen-independent (i.e., decarboxylases) enzymes (fig. 1). The changes in enzymatic activity at high altitude were associated with changes in levels of monoamines and their precursor amino acids: thus, in animals maintained at high altitude throughout development, levels of norepinephrine (NE) and 5-hydroxytryptamine (5-HT) were consistently high in the mesodiencephalon, pons-medulla and corpus striatum, possibly as a consequence of the increased activity of the synthesizing enzymes, as well as increased brain levels of tryptophan in the case of 5-HT (17). Histamine patterns were also affected by high altitude whereby the activity of the synthesizing enzyme, histidine decarboxylase was reduced first in the cerebral cortex and subsequently in the hypothalamus, two areas where histamine is highly concentrated (10).

From these results it can be concluded that prolonged exposure to moderate hypoxia throughout development induces long-lasting effects on the maturation of several neurotransmitter systems. These effects become manifest in the rat at an early age, around 12 days of age, a critical period of development characterized by rapid brain maturation (e.g., active myelinogenesis) and enhanced physiologic competence (e.g., eye opening). With respect to the enzymes for monoamine metabolism, synthesizing enzymes in the brain and in the adrenal medulla are generally more responsive to hypoxia than catabolizing enzymes.

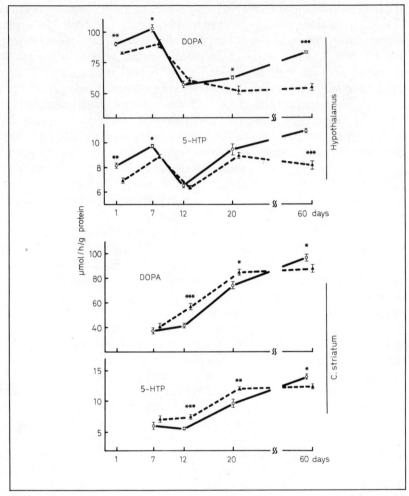

Fig. 1. Effects of exposure to high altitude hypoxia from birth to adulthood on DOPA- and 5-HTP-decarboxylases in several areas of the rat brain. ○ = Sea level, ▲ = high altitude.

Thus, survival and adaptation of the animals born at high altitude occur through modulation of enzymatic activity and the resulting greater availability of monoamines in several brain areas and in the adrenal. Similarly, decreased enzymatic activity in the hypothalamus of rats maturing at high altitude may be associated with a slowing down of monoamine turnover (1) and account for the endocrine imbalances manifested in these rats as stunting of growth, impairment of reproduction and hyperfunction of adrenal cortex (1, 16, 19).

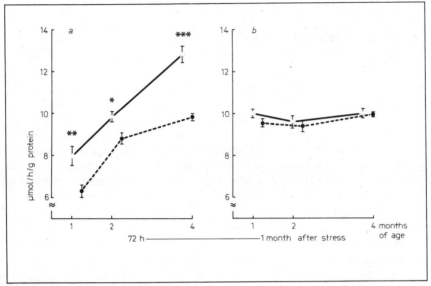

Fig. 2. Effects of combined (noise-light-motion) stress on the development of MAO activity in whole brain. Rats were stressed from 10 days to 4 months of age. Determinations were performed 72 h (a) and 1 month (b) after the last stress. ○ = Controls, ● = stressed.

Noise-Light-Motion Stress

In these experiments, rats were exposed to a stress consisting of a combination of noise (100 dB), flashing lights, and oscillation of the cage, 4 h/day, 4–5 days/week, and MAO activity was measured in whole brain (cerebellum excluded) from 10 days to 4 months of age (9). Under these conditions, several stimuli were operative simultaneously in an effort to better mimic the complexity of the natural environment and to maximize the impact of the environment. The combined stress reduced by 21% MAO specific activity from 10 to 30 days of age (fig. 2) as well as body weight by 10%, effects which persisted when the stress was prolonged for 2 and 4 months. Female rats appeared to be more sensitive to the combined stress than males and showed a greater decrease in MAO activity which persisted for 1 month after cessation of the stress. On the other hand, when the same stress was applied to young-adult and older rats for 1 week only, MAO activity decreased to the same extent (18%) and recovered within the same period in both age groups (8).

The activity of COMT was also measured in these experiments but showed little changes after stress, except for a moderate decrease after 4 months of exposure.

Radiation

Previous studies in the rat have shown that neonatal X-radiation alters the development of the cholinergic system (24, 25) and of γ-aminobutyric acid (23) in various brain regions. Generally, the changes induced are in the direction of an increase in the concentration of these substances. More recent studies have shown that neonatal X-radiation (500 R whole-body) induces profound changes in monoamine metabolism as well (5). NE and 5-HT concentration increased 7 days postradiation in all brain regions undergoing rapid axonal growth and proliferation (e.g., cerebral cortex), but not in the regions of the cell bodies from which the respective neurotransmitter originates (e.g. mesodiencephalon). The increase in NE and 5-HT is accompanied by a concomitant increase in the rate of synthesis. While these changes are evident as late as 22 days of age, the monoaminergic systems revert to normal by maturity. It is suggested that these alterations reflect an imbalance in the density of nerve endings in the region where they terminate.

Because of the preferential destructive action of radiation on dividing cells, radiation has proven to be useful in destroying selected cell populations and thereby isolating specific neurotransmitter systems. An example of the applicability of this property to neurobiologic studies is illustrated by the recent demonstration of glutamic acid as the neurotransmitter of cerebellar granule cells in animals whose cerebellum had been deprived of the granule cell layer either by radiation, infection or mutation (4).

Hypothyroidism

Alterations of thyroid function during development induce structural and biochemical alterations of the central nervous system in humans and experimental animals, including changes in the levels and turnover of several neurotransmitters and their related enzymes. These latter changes are particularly evident in the developing brain where alterations in cholinergic neurotransmission have been reported in homogenates (3) and subcellular compartments (22) and have been associated with impaired synaptogenesis in the hypothyroid brain. Changes in monoamines vary with the neurotransmitter and the brain area in the adult animal (6), and with the age of the animal and the duration of hypothyroidism in the developing animal (21). Thus, neonatal radiothyroidectomy induces significant alterations in metabolic enzymes for monoamines in the cerebral hemispheres, but not in the brain stem, at 13 days of age, and in both regions at 32 days (21). The tendency is for TH activity to increase whereas the other enzyme activities decrease in hypothyroidism. Alterations are also evident in subcellular compartments displaying the highest relative distribution of enzyme activity: for example, TH activity increases in synaptosomes at 13 and 32 days (+ 53 and + 83%, respectively), while at 13 days MAO activity decreases (-13%) in mitochondria and microsomes, and COMT in cytosol (-30%).

In the developing animal, compensatory responses to hypothyroidism do occur and may be mediated through conservation of monoamine levels by increasing their synthesis and/or decreasing their catabolism. Adaptation, however, is not complete, and, if adequate replacement therapy is not initiated at an early critical age, fetal and neonatal hypothyroidism will result in irreversible neurologic and mental deficits such as cretinism in humans. Furthermore, in the adult, although thyroid hormones do not affect such parameters of brain metabolism as oxygen consumption, as they do in the developing animal and in other body tissues, our studies show that these hormones are important for maintenance of neurotransmitter balance (6).

Sex Differences

The role of sex hormones in the differentiation of selected brain areas (e.g., hypothalamus) has been well demonstrated with respect to both functional (e.g., behavioral and endocrine) and biochemical (e.g., receptors for estrogens and androgens) levels. Our studies in the rat show that the expression of these sex differences is also manifested in the development and maintenance of neurotransmission (e.g., monoaminergic pathways), the degree of these differences varying with the brain area and the monoamine considered (18). For example, TH activity is higher in females than in males in cerebral cortex and cerebellum but lower in the pons-medulla and mesodiencephalon. Besides TH, the other synthesizing and catabolizing enzymes are higher in females than in males, and this higher activity is associated with higher NE levels, suggesting that females have more rapid turnover of catecholamines. This appears also to apply to 5-HT in view of the higher activity of the metabolic enzymes and the higher levels of the monoamine in females. In the rat, sex differences in enzyme activity appear at birth and become more marked at puberty, implicating both genetic and hormonal factors.

Conclusions

Within the framework of the questions raised in the Introduction and of the present data, the following general conclusions can be drawn; with respect to developmental ages and to development of specific brain areas, the more rapid the growth and differentiation of the animal and the more immature the brain area, the greater is the susceptibility to the environment. Comparison of the effects induced by environmental and hormonal stimuli suggests that the nature of the response depends on the degree of maturation of the system and the duration and severity of the stimulation rather than the particular type of stimulus. Furthermore, differences are observed among various neurotransmitter systems and, within a given system, individual and regional specificity is apparent. Although even the very young brain is capable of initiating compensatory responses to noxious stimuli, compensation may not be complete and damage may result in irreversible deficits of neurologic functions. On the other hand, the presence of compensatory responses in the adult brain, equally and even more efficient than those taking place in the developing brain, attest to the plasticity of the adult synapse. Such plasticity of neurotransmitter systems has been recently demonstrated also in morphologic (13) and metabolic (7) studies. Among the variables which influence these responses, one which occurs early and persists throughout life is the sex difference. In view of the importance of neurotransmitter substances in synaptic function and control of neural activity and also in view of their role in regulation of endocrine function, alterations in neurotransmission in response to environmental and hormonal stimuli appear to be crucial for the normal establishment of homeostasis.

Summary

The development of cholinergic and monoaminergic systems has been studied in the rat brain under several environmental conditions, both extrinsic (high altitude hypoxia, noise-light-motion stress, radiation) and intrinsic (hypothyroidism, sex differences). The major conclusions to be drawn indicate that the development of neurotransmitter systems in specific brain areas depends on the timetable of maturation of that area – the more immature, the greater its susceptibility; that each neurotransmitter system is differentially affected by the environmental stimulus and varies according to the sex of the animal; and that both developing and adult brain are capable of compensatory responses sufficient to maintain homeostasis.

References

1 *Brotman, S.; Cimino, J.; Vaccari, A.; Umezu, M., and Timiras, P.S.:* Effects of pre- and postnatal hypoxia on the development of brain monoamines and of gonadotropins in the rat; in *Longo* Circulation in the fetus and newborn (Garland Publisher, New York, in press).
2 *Brown, R.M.; Kehr, W., and Carlsson, A.:* Functional and biochemical aspects of catecholamine metabolism in brain under hypoxia. Brain Res. *85:* 491–509 (1975).
3 *Geel, S.E. and Timiras, P.S.:* Influence of neonatal hypothyroidism and of thyroxine on the acetylcholinesterase and cholinesterase activities in the developing central nervous system of the rat. Endocrinology *80:* 1069–1074 (1967).

4 *Hudson, D.B.; Valcana, T.; Bean, G., and Timiras, P.S.:* Glutamic acid: a strong candidate as the neurotransmitter of the cerebellar granule cell. Neurochem. Res. *1:* 73–81 (1976).
5 *Hudson, D.B.; Valcana, T., and Timiras, P.S.:* Monoamine metabolism in the developing rat brain and effects of ionizing radiation. Brain Res. *114:* 571–579 (1976).
6 *Ito, J.M.; Valcana, T., and Timiras, P.S.:* Effect of hypo- and hyperthyroidism on regional monoamine metabolism in the adult rat brain. Neuroendocrinology (in press).
7 *Kaplan, M.S. and Hinds, J.W.:* Neurogenesis in the adult rat: electron microscopic analysis of light radioautographs. Science *197:* 1092–1094 (1977).
8 *Maura, G. and Vaccari, A.:* Relationships between age of submission to environmental stress, and monoamine oxidase activity in rats. Experientia *31:* 191–192 (1975).
9 *Maura, G.; Vaccari, A.; Gemignani, A., and Cugurra, F.:* Development on monoamine oxidase activity after chronic environmental stress in the rat. Environ. Physiol. Biochem. *4:* 64–79 (1974).
10 *Maura, G.; Vaccari, A., and Timiras, P.S.:* Effects of chronic stress on the development of histamine enzymes. Agents Actions *7:* 177–181 (1977).
11 *Petropoulos, E.A. and Timiras, P.S.:* Biological effects of high altitude as related to increased solar radiation, temperature fluctuations and reduced partial pressure of oxygen; in *Tromp* Progress in biometeorology, vol. 1, pp. 295–328; 642–662 (Swets & Zeitlinger, Amsterdam 1974).
12 *Petropoulos, E.A. and Timiras, P.S.:* Effects of hypoxic environment on prenatal brain development: recent evidence versus earlier dogma; in *Vernadakis and Weiner* Drugs and the developing brain, pp. 429–449 (Plenum Publishing, New York 1974).
13 *Scheff, S.; Benardo, L., and Cotman, C.:* Progressive brain damage accelerates axon sprouting in the adult rat. Science *197:* 795–797 (1977).
14 *Timiras, P.S.:* High-altitude studies; in *Gay* Methods of animal experimentation, vol. 2, pp. 333–369 (Academic Press, New York 1965).
15 *Timiras, P.S.:* Developmental changes in the responsivity of the brain to endogenous and exogenous factors; in *Vernadakis and Weiner* Drugs and the developing brain, pp. 417–427 (Plenum Publishing, New York 1974).
16 *Timiras, P.S. and Woolley, D.E.:* Functional and morphologic development of brain and other organs of rats at high altitude. Fed. Proc. Fed. Am. Socs exp. Biol. *25:* 1312–1320 (1966).
17 *Vaccari, A.; Brotman, S.; Cimino, J., and Timiras, P.S.:* Adaptive changes induced by hypoxia in the development of brain monoamine enzymes. Neurochem. Res. (in press).
18 *Vaccari, A.; Brotman, S.; Cimino, J., and Timiras, P.S.:* Sex differentiation of neurotransmitter enzymes in central and peripheral nervous systems. Brain Res. *132:* 176–185 (1977).
19 *Vaccari, A.; Cimino, J.; Brotman, S., and Timiras, P.S.:* High altitude hypoxia and adrenal development in the rat: enzymes for biogenic amines. 27th Int. Congr. Physiol. Sci., Satellite Symp. on Environmental Endocrinology, Montpellier 1977. Proc. in Life Sci. (Springer, Berlin, in press).
20 *Vaccari, A.; Maura, G.; Marchi, M., and Cugurra, F.:* Development of monoamine oxidase in several tissues of the rat. J. Neurochem. *19:* 2453–2457 (1972).
21 *Vaccari, A.; Valcana, T., and Timiras, P.S.:* Effects of hypothyroidism on the enzymes for biogenic amines in the developing rat brain. Pharmacol. Res. Commun. *9:* 763–780 (1977).
22 *Valcana, T.:* Effect of neonatal hypothyroidism on the development of acetylcholinesterase and choline acetyltransferase activities in the rat brain; in *Ford* Influence of hormones on the nervous system, pp. 174–184 (Karger, Basel 1971).

23 Valcana, T.; Hudson, D., and Timiras, P.S.: Effects of X-radiation on amino acid content in the developing rat cerebellum. J. Neurochem. *19:* 2229–2232 (1972).
24 Valcana, T.; Liao, C., and Timiras, P.S.: Effects of X-radiation on the development of the cholinergic system of the rat brain. II. Investigation of alterations in acetylcholine content. Environ. Physiol. Biochem. *4:* 58–63 (1974).
25 Valcana, T. and Timiras, P.S.: Effects of X-radiation on the development of the cholinergic system of the rat brain. I. Study of alterations in choline acetyltransferase and acetylcholinesterase activity and acetylcholinesterase synthesis. Environ. Physiol. Biochem. *4:* 47–57 (1974).
26 Vernadakis, A. and Timiras, P.S.: Disorders of the nervous system; in *Assali* Pathophysiology of gestation, pp. 233–304 (Academic Press, New York 1972).

P.S. Timiras, MD, Department of Physiology-Anatomy, University of California, *Berkeley, CA 94720* (USA)

Maturation of the Responses of Brain 5-Hydroxytryptamine Turnover, Plasma Nonesterified Fatty Acids and Corticosterone to Stress during Ontogeny

Anja H. Tissari and Ilkka T. Tikkanen

Department of Pharmacology, University of Helsinki, Helsinki

Mechanisms controlling transmitter turnover and impulse activity in brain 5-hydroxytryptamine (5-HT) neurons are still insufficiently known. Availability of tryptophan may be one controlling factor because the K_m of tryptophan hydroxylase, the rate-limiting enzyme of 5-HT synthesis, is higher than the brain tryptophan content (17). Brain tryptophan content and 5-HT turnover rate fluctuate parallel in some physiological and pharmacological conditions (6, 22, 26). The content of brain tryptophan appears to depend on the content of plasma free tryptophan which is influenced by changes of the contents of endogenous substances and drugs binding to plasma proteins (15, 26). It is assumed that in fasting and various stress conditions the increase of brain 5-HT turnover is due to the increase of the contents of plasma nonesterified fatty acids (NEFA), free tryptophan and brain tryptophan (15, 25). *Fernstrom and Wurtman* (9) have presented evidence showing that brain tryptophan content depends on the ratio of the plasma tryptophan content to the sum of the plasma contents of other large neutral amino acids competing for the same transport mechanism into the brain (3). The increase of brain 5-HT turnover in stress coincides with the increase of brain tryptophan hydroxylase activity (1, 10), which depends on the increase of plasma glucocorticoids (7) because it is abolished by adrenalectomy (1). Increase of brain 5-HT turnover and tryptophan hydroxylase activity by the administration of reserpine and monoamine oxidase inhibitors (19, 32) is parallel to the increase of the plasma corticosterone content (18, 33).

We have found (27) that during postnatal development in the rat the maturation of the responses of brain 5-HT turnover to fasting and immobilization stress has a totally different time course. Fasting causes a much higher increase of brain 5-HT turnover in neonatal than adult rats. On the contrary, immobilization does not have any effect in neonatal rats before the age of 3

Table I. Effect of fasting on the contents of brain tryptophan, 5-HIAA and 5-HT and plasma tryptophan and NEFA in rats

Age and treatment	Brain			Plasma		
	tryptophan µg/g	5-HIAA ng/g	5-HT ng/g	tryptophan, µg/ml		NEFA µEq/ 1,000 ml
				total	free	
3 h						
0 h control	19.6 ± 1.4	401 ± 26	234 ± 7			
16 h control, %	49.2	82.6	109			
16 h fasting, %	101**	121**	138			
1 day						
0 h control	5.3 ± 0.4	232 ± 14	184 ± 8	11.0 ± 1.5	5.7 ± 0.8	196 ± 21
16 h control, %	158	128	106	146	147	88.9
16 h fasting, %	475***	266***	155***	243**	249*	91.0
1 week						
16 h control	6.9 ± 0.4	353 ± 8	243 ± 13	23.2 ± 2.0	6.9 ± 1.2	321 ± 16
16 h fasting, %	239***	161***	129***	81.6	161*	173***
Adult						
24 h control	4.8 ± 0.3	485 ± 30	510 ± 26	23.5 ± 0.9	1.0 ± 0.1	253 ± 19
24 h fasting, %	115	120*	107	108	144*	188***

From the same brain samples tryptophan (8), 5-HIAA and 5-HT (5) were analyzed. Plasma free tryptophan was measured by ultrafiltration using Amicon Centriflo 50 dialysis cones and NEFA by the method of *Novak* (21). The effect of staying in normal and/or fasting conditions is expressed as percent of the control content given in the table. Mean ± SEM of 6 experiments. * = $p < 0.05$, ** = $p < 0.01$, *** = $p < 0.001$ when compared with the simultaneous control.

weeks. These observations prompted us to examine in postanatal rats relationships between the maturation of the responses of brain 5-HT turnover and of plasma NEFA and corticosterone to some physiological and pharmacological stimuli. Results concerning the effects of fasting and immobilization stress (28) are summarized in this paper.

Table I shows the effect of fasting on the contents of plasma NEFA, total and free tryptophan and brain tryptophan, 5-hydroxyindoleacetic acid (5-HIAA) and 5-HT. The measurements were made after 16 h fasting in 3-hour, 1-day and 1-week-old rats; at these ages the increase of brain 5-HT turnover is remarkably greater than that in adults. The adult contents were determined after 24 h fasting. In 3-hour-old rats fasting prevented the 50% decrease of the brain tryptophan content occurring in control conditions and increased the brain 5-HIAA and 5-HT contents by 50 and 30%, respectively. In 1-day-old rats fasting did not cause any change in the content of plasma NEFA but increased those of total and free tryptophan to 170% of controls, probably due to decrease of protein synthesis. The content of brain tryptophan increased to 300, that of 5-HIAA to 210 and 5-HT to 150% of controls. In 1-week-old rats the content of plasma NEFA increased to 170% of controls and that of free tryptophan increased parallel. The contents of brain tryptophan, 5-HIAA and 5-HT rose to 240, 160 and 130% of controls, respectively. In adult rats the increases of plasma NEFA and free tryptophan contents were roughly similar to those in 1-week olds but the contents of brain tryptophan and 5-HIAA rose only to 115 and 120% of controls, respectively. As a comparison, plasma and brain tyrosine levels were measured and the only significant change detected was a decrease of plasma tyrosine level in 1-day and 1-week-old rats to 72 and 53% of controls, respectively. Thus, though during fasting the proportional increase of plasma free tryptophan content was roughly similar in all age groups, the proportional increases of brain tryptophan, 5-HIAA and 5-HT contents were more than 10 times greater in neonatal than adult rats.

Vahvelainen and Oja (30) have demonstrated that the transport system for large neutral amino acids prefers phenylalanine over the other amino acids to a greater extent in immature than adult brain. Thus phenylalanine has a higher inhibitory effect on tryptophan transport and the transport of phenylalanine is less inhibited by other amino acids in slices from 7-day-old than from adult rat brain. In neonatal rats during fasting the phenylalanine contents of plasma and brain may have decreased and the inhibition of tryptophan transport by phenylalanine may have diminished causing high increase of the brain tryptophan content detected in the present study. A lack of phenylalanine is suggested by the decreased plasma tyrosine levels found in fasted neonatal rats. An additional contributing factor may have been the higher affinity of the transport for tryptophan in neonatal than in adult rats (4, 29).

To assess further the above assumption, *l*-phenylalanine and *l*-tyrosine were administered to 1-day-old rats (1,000 mg/kg i.p. 3 times, at 0, 6 and 12 h, animals were killed at 16 h) (fig. 1). Phenylalanine treatment prevented nearly completely the increase of brain tryptophan content during fasting but did not have any effect in control conditions. In addition, in both test conditions the contents of brain 5-HIAA and 5-HT decreased more than that of tryptophan. Thus phenylalanine may have inhibited both tryptophan transport and hydroxylation. *Lovenberg et al.* (17) have shown that phenylalanine is a competitive inhibitor and possibly a substrate for brain tryptophan hydroxylase. Tyrosine treatment did not affect the increase of brain tryptophan during fasting but unexpectedly increased it in control conditions to 230% of the saline control. The latter effect was evidently due to fasting since tyrosine was given as a suspension which decreased peristalsis and the rats stopped to eat. After tyrosine treatment in both test conditions the increases of the contents of brain 5-HIAA

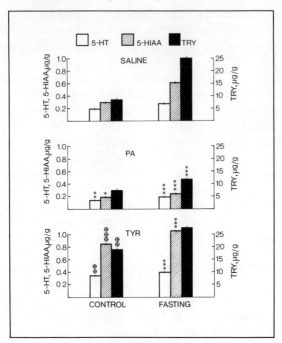

Fig. 1. Effects of phenylalanine and tyrosine on the contents of brain tryptophan, 5-HIAA and 5-HT in 1-day-old rats. L-Phenylalanine (PA) and l-tyrosine (TYR) were administered in control and fasting conditions 1,000 mg/kg i.p. 3 times, at 0, 6 and 12 h and the animals were killed at 16 h. Measurements were made as described in table I. * = p <0.05, ** = p <0.01, *** = p <0.001 when compared with the respective saline control; ⊕⊕ = p <0.01, ⊕⊕⊕ = p <0.001 when compared with the fasted saline control.

and 5-HT were markedly greater than that of tryptophan. This effect may have been caused by the high tyrosine content attained in the brain, possibly also in 5-HT neurons, which may have decreased the affinity of tryptophan hydroxylase for phenylalanine and diminished the inhibition of tryptophan hydroxylation by phenylalanine. *Karobath and Baldessarini* (14) have demonstrated in synaptosomes that tyrosine is an effective inhibitor of hydroxylation of phenylalanine. The formation of catecholamines only was measured in their study so it is not known whether hydroxylation of phenylalanine occurred also in 5-HT terminals. The results from the administration of these two amino acids support the assumption that the high increase of brain 5-HT turnover in neonatal rats during fasting is due to a lack of phenylalanine and occurs at the steps of tryptophan transport and hydroxylation.

Parallel to our results, high brain tryptophan contents have been observed in undernourished 4-day-old rats (4). The capacity to synthesize 5-HT from trypto-

Table II. Effect of immobilization on the contents of brain tryptophan, 5-HIAA and 5-HT and plasma tryptophan and NEFA in rats

Age and treatment	Brain			Plasma		
	tryptophan µg/g	5-HIAA ng/g	5-HT ng/g	tryptophan, µg/ml		NEFA µEq/ 1,000 ml
				total	free	
1 day						
control	7.9 ± 1.2	340 ± 24	256 ± 20	18.3 ± 2.3	9.5 ± 1.9	280 ± 10
isolation, %	134	104	99.0	117	102	75.5
immobilization, %	136	99.3	111	94.0	132	94.2
3 weeks						
control	5.7 ± 0.4	556 ± 31	338 ± 18	24.9 ± 1.1	2.1 ± 0.1	340 ± 27
isolation, %	109	95.5	106	89.2	85.7	106
immobilization, %	114	128***	103	80.0	78.3	104
Adult						
control	4.1 ± 0.2	508 ± 20	553 ± 30	20.7 ± 0.6	1.1 ± 0.1	212 ± 13
immobilization, %	116**	142***	105	74.1***	130	189***

Rats were isolated from the mother or immobilized for 4 h. The effects are expressed as percent of the control content given in the table. Measurements were made as described in table I. Mean ± SEM of 6 experiments. ** = p <0.01, *** = p <0.001 when compared with the isolated animal (postnatals) or with the control (adults).

phan is present in the rat brain already on the 15th fetal day (12). In the present study the increase of 5-HT turnover by fasting was similar in brain stem and hemispheres in 1-day-old, rats, when brain monoamine terminals are very immature (13).

To examine mechanisms related to immobilization stress, the changes of the contents of plasma NEFA, total and free tryptophan and brain tryptophan, 5-HIAA and 5-HT were studied. The measurements were made after 4 h immobilization in 1-day-old rats when brain 5-HT turnover does not yet respond, in 3-week-old rats when a significant increase is first obtained, and in adults (table II). In 1-day-old rats immobilization did not cause any change in the plasma and brain contents. In 3-week-old rats any of the plasma contents and that of brain tryptophan did not change but the content of brain 5-HIAA increased significantly to 135% of the control. In adult rats immobilization increased the content of plasma NEFA to 190% and that of free tryptophan to 130% of controls while the total tryptophan content decreased. The content of brain tryptophan increased slightly but significantly and that of brain 5-HIAA

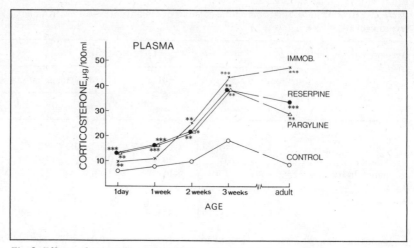

Fig. 2. Effects of immobilization and the administration of reserpine and pargyline on the plasma corticosterone content. Measurements (31) were made after 1 h immobilization and 2 h after the i.p. administration of reserpine (4 mg/kg to postnatal and 5 mg/kg to adult rats) and pargyline (100 mg/kg). * = $p < 0.05$, ** = $p < 0.01$, *** = $p < 0.001$ when compared with the control.

rose to 142% of controls. In another series of experiments in adult rats, the increase of plasma NEFA content in 2 h immobilization stress was prevented by the administration of atenelol, a β-adrenergic blocking agent (10 mg/kg i.p.). The brain 5-HIAA content increased then as highly as in the control rats. Thus neither in 3-week-old nor in adult rats the increase of brain 5-HT turnover was secondary to the increase of the plasma NEFA content.

The responses of plasma and adrenal corticosterone levels to immobilization were studied after 1 h stress, the maximum effect in adult rats is obtained after 30 min (20). The plasma corticosterone level (fig. 2) increased significantly by 40% in 1-day and by 140% in 2-week-old rats. In 3-week-old rats the adult level after stress was attained though the proportional increase was only 150 versus 500% in adults. As a comparison, the effects of reserpine (4 mg/kg to postnatal and 5 mg/kg to adult rats i.p.) and pargyline, a monoamine oxidase inhibitor (100 mg/kg i.p.) were studied. The animals were killed 2 h after the drug administration. The increase of plasma corticosterone content caused by these drugs was greater than that caused by immobilization in 1-day and 1-week olds and similar or smaller than that in the older rats. The changes of adrenal corticosterone paralleled those of plasma except that immobilization in 1-day and 1-week-old rats caused a significantly higher increase of the adrenal than the plasma corticosterone. Similar changes of plasma and adrenal corticosterone contents of neonatal rats after different stress conditions have been reported earlier (11).

The increase of plasma corticosterone by immobilization appeared earlier during postnatal ontogeny than that of brain 5-HT turnover. A similar discrepancy was observed in the effect of reserpine. Though reserpine caused in 1-day-old rats a significant increase of the plasma corticosterone content, it did not increase the turnover of brain 5-HT (mean ± SEM of the brain 5-HIAA contents, $\mu g/g$, 1-day olds: saline control 0.31 ± 0.02, saline plus 12 h fasting 0.56 ± 0.03, reserpine 4 mg/kg plus 12 h fasting 0.53 ± 0.03; adults: saline 0.52, reserpine 5 mg/kg 12 h earlier 1.3 ± 0.1).

Sze et al. (24) have recently reported that the contents of brain corticosterone and tryptophan hydroxylase activity start to increase simultaneously on the 6th postnatal day in the rat. The developmental increase of brain tryptophan hydroxylase activity is prevented by adrenalectomy and restored to a higher than normal level by a continued corticosterone treatment. Between postnatal days 1 and 5 tryptophan hydroxylase of the rat brain does not respond to glucocorticoids (23). In the experiments of *Sze et al.* (24) the increase of the enzyme protein by the prolonged action of corticosterone was measured. In an acute stress the elevation of tryptophan hydroxylase activity is not due to the increase of the enzyme synthesis but the increase of the catalytic efficiency of the existing enzyme (1). Corticosterone injection causes an early (1 h) increase of tryptophan hydroxylase activity unrelated to the enzyme synthesis in forebrain and midbrain whereas the later (4 h) increase in midbrain results from the increased synthesis of enzyme protein (2). The earliest and greatest increases of the brain 5-HIAA content in immobilization stress occur in the cerebral cortex (20). Possibly for the increase of brain 5-HT turnover in stress the function of 5-HT nerve terminals is essential because this response appears at a time when these neurons reach the adult stage morphologically (16).

Summary

Fasting caused a much higher increase of brain 5-HT turnover in neonatal than adult rats whereas immobilization had no effect before the age of 3 weeks. 16 h fasting increased in 1-day-old rats the contents of brain tryptophan, 5-HIAA and 5-HT to 300, 210 and 150% and those of plasma total and free tryptophan to 170% of controls, respectively, the content of plasma NEFA did not rise before the age of 1 week. The increases were prevented by the administration of phenylalanine during fasting. 4 h immobilization increased in 3-week-old rats the content of brain 5-HIAA to 135% of the control without affecting the contents of brain tryptophan, plasma NEFA and free tryptophan. A significant increase of plasma corticosterone content was found already in 1-day-old rats. These responses in neonatal rats were different from those in adults in which fasting and immobilization increased by 90% the content of plasma NEFA, significantly that of plasma free tryptophan, slightly the brain tryptophan content and by 20 and 42%, respectively, the content of brain 5-HIAA.

References

1 *Azmitia, E.C., jr. and McEwen, B.S.:* Adrenalcortical influence on rat brain tryptophan hydroxylase activity. Brain Res. *78:* 291–302 (1974).
2 *Azmitia, E.C., jr. and McEwen, B.S.:* Early response of rat brain tryptophan hydroxylase activity to cycloheximide, puromycin and corticosterone. J. Neurochem. *27:* 773–778 (1976).
3 *Blasberg, R. and Lajtha, A.:* Substrate specificity of steady-state amino acid transport in mouse brain slices. Archs Biochem. Biophys. *112:* 361–377 (1965).
4 *Bourgoin, S.; Faivre-Bauman, A.; Benda, P.; Glowinski, J., and Hamon, M.:* Plasma tryptophan and 5-HT metabolism in the CNS of the newborn rat. J. Neurochem. *23:* 319–327 (1974).
5 *Curzon, G. and Green, A.R.:* Rapid method for the determination of 5-hydroxytryptamine and 5-hydroxyindoleacetic acid in small regions of rat brain. Br. J. Pharmacol. *69:* 653–655 (1970).
6 *Curzon, G.; Joseph, M.H., and Knott, P.J.:* Effects of immobilization and food deprivation on rat brain tryptophan metabolism. J. Neurochem. *19:* 1967–1974 (1972).
7 *De Schaepdryver, A.; Preziosi, P., and Scapagnini, U.:* Brain monoamines and adrenocortical activation. Br. J. Pharmacol. *35:* 460–467 (1969).
8 *Denckla, W.D. and Dewey, H.K.:* The determination of tryptophan in plasma, liver and urine. J. Lab. clin. Med. *69:* 160–169 (1967).
9 *Fernstrom, J.D. and Wurtman, R.J.:* Brain serotonin content. Physiological regulation by plasma neutral amino acids. Science, N.Y. *178:* 414–416 (1972).
10 *Gal, E.M.; Heater, R.D., and Millard, S.A.:* Studies on the metabolism of 5-hydroxytryptamine (serotonin). VI. Hydroxylation and amines in cold-stressed reserpinized rats. Proc. Soc. exp. Biol. Med. *128:* 412–415 (1968).
11 *Haltmeyer, G.C.; Denenberg, V.H.; Thatcher, J., and Zarrow, M.X.:* Response of the adrenal cortex of the neonatal rat after subjection to stress. Nature, Lond. *212:* 1371–1373 (1966).
12 *Howd, R.A.; Nelson, M.F., and Lytle, L.D.:* L-Tryptophan and rat fetal brain serotonin. Life Sci. *17:* 801–812 (1975).
13 *Kanerva, L.; Tissari, A.H.; Suurhasko, B.V.A., and Hervonen, A.:* Ultrastructural characterization of synaptosomes from neonatal and adult rats with special reference to monoamines. J. comp. Neurol. *174:* 631–658 (1977).
14 *Karobath, M. and Baldessarini, R.J.:* Formation of catechol compounds from phenylalanine and tyrosine with isolated nerve endings. Nature new Biol. *237:* 57–58 (1972).
15 *Knott, P.J. and Curzon, G.:* Free tryptophan in plasma and brain tryptophan metabolism. Nature, Lond. *239:* 452–453 (1972).
16 *Loizou, L.:* The postnatal ontogeny of monoamine-containing neurones in the central nervous system of the albino rat. Brain Res. *40:* 395–418 (1972).
17 *Lovenberg, W.; Jequir, E., and Sjoerdsma, A.:* Tryptophan hydroxylation in mammalian systems. Adv. Pharmacol. *6A:* 21–36 (1968).
18 *Martel, R.R.; Westermann, E.O., and Maickel, R.P.:* Dissociation of reserpine-induced sedation and ACTH hypersecretion. Life Sci. *1:* 151–155 (1962).
19 *Millard, S.A. and Gal, E.M.:* The contribution of 5-hydroxyindolepyruvic acid to cerebral 5-hydroxyindole metabolism. Int. J. Neurosci. *1:* 211–218 (1971).
20 *Morgan, W.W.; Rudeen, P.K., and Pfeil, K.A.:* Effect of immobilization stress on serotonin content and turnover in regions of the rat brain. Life Sci. *17:* 143–150 (1975).

21 *Novak, M.:* Colorimetric ultramicro-method for the determination of free fatty acids. J. Lipid Res. *6:* 431–433 (1965).
22 *Perez-Cruet, J.; Tagliamonte, A.; Tagliamonte, P., and Gessa, G.L.:* Changes in brain serotonin metabolism associated with fasting and satiation in rats. Life Sci. *11:* 31–39 (1972).
23 *Sze, P.Y.:* Glucocorticoid regulation of the serotonergic system of the brain; in Costa, Giacobini and Paoletti First and second messengers: new vistas. Advances in biochemical psychopharmacology, vol. 15, pp. 251–265 (Raven Press, New York 1976).
24 *Sze, P.Y.; Neckers, L., and Towle, A.C.:* Glucocorticoids as a regulatory factor for brain tryptophan hydroxylase. J. Neurochem. *26:* 169–173 (1976).
25 *Tagliamonte, A.; Biggio, G.; Vargiu, L., and Gessa, G.L.:* Free tryptophan in serum controls brain tryptophan level and serotonin synthesis. Life Sci. *12:* 277–287 (1973).
26 *Tagliamonte, A.; Tagliamonte, P.; Perez-Cruet, J.; Stern, S., and Gessa, G.L.:* Effect of psychotrpic drugs on tryptophan concentration in the rat brain. J. Pharmac. exp. Ther. *177:* 475–480 (1971).
27 *Tissari, A.H.:* Pharmacological and ultrastructural maturation of serotonergic synapses during ontogeny. Med. Biol. *53:* 1–14 (1975).
28 *Tissari, A.H. and Tikkanen, I.T.:* Mechanisms controlling turnover of brain 5-HT during fasting and immobilization in postanatal rats. Abstr. 6th Int. Congr. of Pharmacology, Helsinki 1975, p. 520.
29 *Vahvelainen, M.-L. and Oja, S.S.:* Kinetics of the influx of phenylalanine, tyrosine, tryptophan, histidine and leucine into slices of brain cortex from adult and 7-day-old rats. Brain Res. *40:* 477–488 (1972).
30 *Vahvelainen, M.-L. and Oja, S.S.:* Kinetic analysis of phenylalanine-induced inhibition in the saturable influx of tyrosine, tryptophan, leucine and histidine into brain cortex slices from adult and 7-day-old rats. J. Neurochem. *24:* 885–892 (1975).
31 *Zenker, N. and Bernstein, D.E.:* The estimation of small amounts of corticosterone in rat plasma. J. biol. Chem. *231:* 695–701 (1958).
32 *Zivkovic, B.; Guidotti, A., and Costa, E.:* Increase of tryptophan hydroxylase activity elicited by reserpine. Brain Res. *57:* 522–526 (1973).
33 *Zivkovic, B.; Guidotti, A., and Costa, E.:* On the regulation of tryptophan hydroxylase in brain; in Costa, Gessa and Sandler Serotonin: new vistas. Advances in biochemical psychopharmacology, vol. 11, pp. 19–30 (Raven Press, New York 1974).

A.H. Tissari, MD, Department of Pharmacology, University of Helsinki, Siltavuorenpenger 10, *SF–00170 Helsinki 17* (Finland)

Development of Biochemical Correlates of Behavior

Is Prenatal Induction of Tyrosine Hydroxylase Associated with Postnatal Behavioral Changes?[1]

S.B. Sparber

Departments of Pharmacology and Psychiatry, University of Minnesota, Minneapolis, Minn.

During this past decade there has been a resurgence of interest in developmental phenomena, including the disciplines within the neurosciences. Unlike the emphasis upon gross structure and function, which took place earlier, more recent interest evolves around biochemical and metabolic activity of the peripheral and central nervous system, as they relate to normal and abnormal development (this symposium, 17, 28). Experimental psychologists interested in development of behavior have been studying effects of various environmental manipulations, such as stress or simple handling procedures upon maternal-offspring interaction during the perinatal period and behavior of offspring in later life (16). There are ample demonstrations of these so-called early experience phenomena to question the validity of developmental behavioral toxicological or behavioral teratological studies which purport to show direct effects of drugs or chemicals upon behavior or any other measure of CNS function in later life, without including postnatal fostering procedures to account, at least partially, for these variables (7). Also, recent studies demonstrating altered nesting and other forms of maternal behavior resulting from stress to rat pups during the early postnatal period (4) reinforce the idea that even transient drug effects upon the behavior of mothers, offspring or both during the preweaning stages of development may produce behavioral changes in adulthood which can be misconstrued as a behavioral teratogenic (29) outcome. Therefore, even fostering procedures may not control for factors in which treated pups (e.g., *in utero* exposure) are given to nontreated foster mothers who respond 'abnormally' because the pups are behaving 'abnormally'. The 'abnormal' behavior of the pups could be due to direct action of a psychoactive drug with a long half-life or because they are experiencing withdrawal from a

---

[1] Supported in part by USPHS grant MH08565 and Minnesota Medical Foundation grant FSW-1-71.

dependence-producing agent administered to their biological mothers during pregnancy. (See *Bignami*, 5, for an excellent review of the recent literature on developmental behavioral toxicology.)

Biochemical-physiological events which covary or correlate well with behavior may not be the substrates of that behavior. As such, the behavioral variable may be the antecedent and responsible for observed changes in the biological variables (18, 21). Therefore, by extension of the early experience, fostering, etc. factors introduced above, a demonstration of biochemical-physiological changes in adulthood, of animals exposed to drugs, chemicals or stress *in utero* or during early postnatal development, likewise is not sufficient evidence demonstrating a *direct* action of insult upon the developing organism.

The relatively few behavioral teratology studies, most of which do not control for the variables cited above, imply that biochemical-pharmacological factors are responsible for behavior changes reported. Additionally, although most drugs (which are lipophilic) can cross the placenta, few studies demonstrated that the organism was actually exposed to the agent early in development. This factor is important for two reasons. Firstly, in the case where a positive outcome (postnatal behavioral effects) is observed, the agent may be affecting the maternal milieu without exposing the embryo or fetus and therefore an indirect action may be responsible (30). Secondly, from a societal point of view, the risk potential to the developing human fetus may be underestimated from infrahuman studies in which no apparent or detectable effects are demonstrated. The drug may not have been getting to the embryo and/or having a pharmacological-toxicological action because of pharmacokinetic or dispositional factors (15).

In order to be able to answer some of the questions relating to factors cited above, in addition to more basic questions relating to control mechanisms of transmitter biosynthesis and storage and how these might interact with behavior development, we have been studying some of these variables in the domestic chicken. This species allows us to administer drugs or chemicals directly into the egg, almost assuring exposure during development. Other advantages to using this model, including the absence of maternal variables, have been described elsewhere (20).

To verify that manipulation of biogenic amines could be effected during embryogenesis by injection of reserpine into eggs, we analyzed chick embryo brain for catecholamines. Doses of reserpine which had little or no effect upon hatchability when injected into eggs prior to incubation (e.g., 0.1 mg/kg egg) produced about a 50% decrease in embryonic brain catecholamines at 14 and 18 days of development (24). Additionally, within a day after hatching, catecholamines were still depleted in their brains and the capacity to take up and store exogenous norepinephrine (^{3}H-NE) administered 30 min prior to sacrifice, was also impaired (25).

The first indication that behavioral development might be impaired or modified was the observation that kicking by embryos subsequent to decapitation for brain analyses was significantly depressed at times when brain catecholamines were depleted (20). That the reduced kicking may have had some functional significance, vis-à-vis normal fixed-action patterns, was indicated by a significant shift or delay in hatching by chicks exposed to the drug during embryogenesis (22). Continued behavioral and biochemical analyses of these chickens during the 30 days after hatching indicated that several classes of behavior, including imprinting-following behavior, detour learning, conditioned avoidance behavior and locomotor activity were all affected by prenatal exposure to reserpine. However, exposure to reserpine, beginning at different stages of development (e.g., prior to incubation, after 7 days of embryogenesis or after 15 days of embryogenesis) resulted in different behavioral consequences, ranging from no effect to an opposite effect. These behavioral differences were not only time-dependent but dose-dependent, indicating that residual, acute pharmacological effects of the drug administered late in development may mask or counteract any longer-lasting developmental action (22).

When we observed small but significant elevations in protein content of embryonic brain and a significant increase in steady-state levels of catecholamines in 30-day-old chicks hatched from reserpinized eggs, we postulated that a critical period during development existed in which permanent changes in ontogenetic expression of transmitter control mechanisms could be permanently affected by drugs (25). This was consistent with a similar suggestion made by *Eiduson* (9) but for which data were lacking.

In a series of experiments designed to determine if induction of tyrosine hydroxylase may be responsible for or associated with the biochemical and/or behavioral outcome of depletion of catecholamines during development, we examined the effects of reserpine upon enzyme activity during embryogenesis, soon after and long after hatching. Tyrosine hydroxylase specific activity in 10-day-old chick embryo brain is about 2% of the activity in the 'mature' brain (29-day-old postnatally, 40 days after sampling the 10-day-old embryo). Between this stage of development and day 20 (1 day prior to hatching) there is a 33-fold increase so that about 70% of the activity measured 29 days postnatally is present at the time of hatching (12). These rates of development are remarkably close to those reported by *Coyle and Axelrod* (8) for the more primitive regions of rat brain (medulla-pons and midbrain-hypothalamus), which would be the appropriate anatomical analog for the cortex-sparse chick. The major difference between the two species seems to be the time of appearance of measurable quantities of the enzyme, about 10 days in the chick embryo and 15 days in the rat embryo.

Examination of 10-day-old embryonic brain indicated almost complete absence of catecholamines (about 5% of controls) in reserpinized embryos. As previously observed, depletion of catecholamines persisted through at least the third day postnatally. When examined 29 days after hatching we again observed a significant increase in steady-state amine levels (about 150% of control). Although catecholamines were severely depleted at 10 days of embryogenesis, tyrosine hydroxylase activity was not different from controls at this time. However, by 15 days of embryogenesis, specific activity of this enzyme was significantly elevated and remained so throughout embryogenesis, 3 days postnatally when catecholamines were still depleted and most interestingly, at 29 days postnatally, when catecholamine levels were 50% above control values (12). *Peters et al.* (18) have reported electrical activity in 14-day-old chick embryo brain but not earlier. It therefore seems reasonable to interpret our finding that increased enzymic activity at 15 days of embryogenesis is consonant with the idea that increased neuronal activity is necessary for induction to occur (27). That the effect was specific for the rate-limiting enzyme was indicated by the fact that dopa decarboxylase was unaltered at any time.

The possibility that a critical period existed for the effect upon tyrosine hydroxylase was further indicated by our finding that treatment at 7 or 14 days of embryogenesis, while depleting catecholamines during the perinatal period, did not increase steady-state levels at 29 days postnatally nor was tyrosine hydroxylase activity altered at this time (fig. 1).

Because of the general lack of specificity of reserpine upon storage vesicles and ultimately levels of monoamines (3), it was of interest to determine if a more selective depletor of catecholamines during early embryonic development would likewise increase tyrosine hydroxylase activity postnatally. This would support our conjecture that the apparent permanent effect on control mechanisms was due to reduction of product during a critical period. α-Methyl-p-tyrosine methyl ester was administered to chicken eggs prior to incubation. While brain catecholamines were depleted to 50% of controls at 10 days of embryogenesis, repletion was complete 1 day prior to hatching (day 20). Maximal depletion was achieved with 10 mg/kg egg, 33 mg/kg egg killing many more embryos but not further depleting brain catecholamines. When whole brain tyrosine hydroxylase, dopa decarboxylase and catecholamines were assayed 29 days postnatally, only tyrosine hydroxylase was significantly elevated. Brain regions (brain stem and optic lobes-cerebral hemispheres) showed dose-dependent increases in this enzyme's activity, even though a 10-fold difference between these regions' basal activities existed (fig. 2). Because we did not observe increased steady-state levels conjointly with enzyme activity increases, we thought that the reserpine treatment altered (increased) the capacity to store product in addition to the increased capacity to synthesize it. As such, crude synaptosomal preparations from the brains of the 29-day-old chickens were studied for their capacity to take up and/or store exogenous catecholamines. No differences were observed between controls, reserpine treated or α-methyl-p-

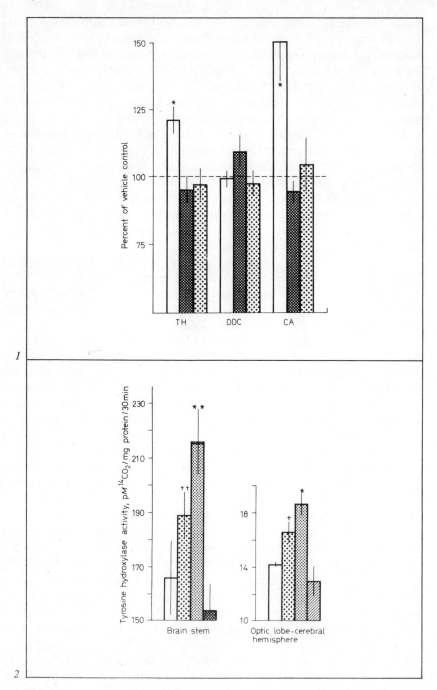

tyrosine treated birds with respect to synaptosomal uptake of ^{3}H-NE without drugs in the medium or with cocaine (3×10^{-5} M) or reserpine (10^{-6} M) in the medium. This rather gross estimate of uptake and storage capacity failed to demonstrate a difference between the two treatments (11).

However, when chicks hatched from eggs injected with α-methyl-p-tyrosine were examined for their unconditioned behavioral and conditioned behavioral characteristics, apparent differences between the two treatments began to emerge (13). Various unconditioned behaviors in the open-field failed to indicate congenital differences between controls and treated birds. Acquisition of the detour response, which showed reserpine-treated birds to be retarded in their rate of learning this task (23), was enhanced by treatment with α-methyl-p-tyrosine.

To verify that a congenital difference in conditioned behavior acquisition existed as a consequence of prenatal treatment with this agent, 55-day-old chickens were allowed access to food in an autoshaping procedure (6). A significant difference due to treatment with the high dose of α-methyl-p-tyrosine emerged during the succeeding five sessions (fig. 3).

While we have not studied autoshaped key-pecking behavior in birds hatched from reserpinized eggs, we have studied both detour learning and

---

Fig. 1. Effects of injection of reserpine (0.17 mg/kg of egg) into the yolk sac at various times during development on whole brain tyrosine hydroxylase (TH) and dopa decarboxylase (DDC) specific activities and whole brain catecholamine (CA) levels at 29 days postnatally. Vertical bars represent mean (expressed as percentage of vehicle control) ± SEM. Six animals were used per group for enzyme assays and for CA determinations. Open bars indicate injection prior to incubation. Cross-hatched and dotted bars indicate injection at 7 and 14 days of embryogenesis, respectively. Control TH specific activities (100%) were 283 ± 12, 385 ± 4 pmol/mg protein/30 min for respective times of injection. DDC specific activities for appropriate controls were 19.40 ± 1.60, 18.75 ± 1.78 and 19.57 ± 1.16 nmol/mg protein/30 min. CA levels for the appropriate controls were 0.26 ± 0.01, 0.32 ± 0.02 and 0.28 ± 0.02 μg/g of brain. * = p <0.01, reserpine vs. vehicle control. Reprinted with permission from *Lydiard and Sparber* (12).

Fig. 2. Effect of injection of α-methyl-p-tyrosine (AMPT) (10 and 33.3 mg methyl ester/kg of egg) prior to incubation on brain stem and optic lobes-cerebral hemispheres TH specific activity at 29 days postnatally. Vertical bars represent mean ± SEM of 7 observations for brain stem and 15 observations for optic lobes-cerebral hemispheres. Open bar represents vehicle treatment only. Dotted and shaded bars represent 10 and 33.3 mg AMPT/kg, respectively. In order to control for a delayed absorption effect, the maximum dose any chick would be exposed to (2 mg/chick) was injected within 24 h after hatching and is referred to as the absolute dose. TH activity of this treatment is represented by the cross-hatched bars (n = 8, optic lobes-cerebral hemispheres). † = p <0.05; †† = p <0.025; * = p <0.01; ** = p <0.005; AMPT vs. vehicle-injected control, one-tailed t-test after one-way analysis of variance. For brain stem F(3,24) = 5.7011; for optic lobes-cerebral hemispheres F(2,44) = 10.0096. Reprinted with permission from *Lydiard et al.* (11).

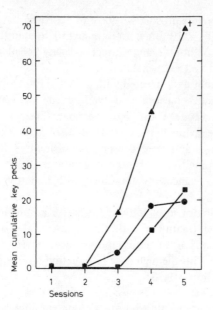

Fig. 3. Effect of prenatal administration of 10.0 and 33.3 mg AMPT/kg of egg, prior to incubation, on mean cumulative key peck responses in 55- to 60-day-old chickens. Circles, squares, and triangles represent groups on vehicle (saline), 10.0, and 33.3 mg AMPT/kg of egg, respectively (8 animals in each group). † = $p < 0.005$. Reprinted with permission from *Lydiard and Sparber* (13).

autoshaped key-pecking in birds hatched from eggs injected with the environmental pollutant, methylmercury. With this organic mercurial, we observed impaired detour learning when injected early during embryogenesis (prior to incubation) but not later (day 7 or day 14 of embryogenesis), even though the older embryos were more sensitive to the killing effects of the chemical (10).

Since hatching behavior of the methylmercury-exposed chicks was retarded (10) in a manner similar to that produced by reserpine (22), one might infer that other behavioral variables would show similar profiles for mercury and reserpine groups.

As such, given the license to extrapolate from exposure to methylmercury to that of reserpine (e.g., α-methyl-*p*-tyrosine ≠ reserpine, reserpine = methylmercury ∴ α-methyl-*p*-tyrosine ≠ methylmercury), this inductive logical manipulation would predict that methylmercury-exposed chicks would show retarded acquisition of the autoshaped key-pecking operant, since the α-methyl-*p*-tyrosine group showed enhanced acquisition. When tested in this paradigm, that was the outcome (*Sparber and Rosenthal*, unpublished observations).

In conclusion, we have demonstrated that drugs can directly alter biochemical and rudimentary and mundane behavioral parameters (postdecapitation kicking, hatching behavior) prior to hatching when injected into the eggs of domestic chickens. We have also demonstrated that critical periods exist for biochemical and behavioral teratogenic actions of drugs and chemicals in this species. Finally, we have demonstrated that similar increases in specific activity of tyrosine hydroxylase in brains of chicks exposed to catecholamine-depleting procedures prior to hatching may be associated with opposite behavioral effects postnatally; one of the differentiating biochemical characteristics being differences in steady-state levels of monoamines in their brains postnatally. While the relevance of developmental studies utilizing *Gallus domesticus* to problems of human development is equivocal, studies of the congenital behavioral effects of methylmercury in rodents and chicks (10, 26) showing effects analogous to human fetal exposure to methylmercury (1, 14) lend credence to the idea that a useful biological continuum exists across the various vertebrate species.

Confirmation that reserpine exposure during development of mammalian species can lead to permanent changes in tyrosine hydroxylase activity postnatally was offered by *Bartolomé et al.* (2). While the direction of change in enzyme activity of rat adrenals postnatally, subsequent to treatment of pregnant mothers 9–7 days prior to term, was the same as seen in early exposed chick brain, rat brain tyrosine hydroxylase was suppressed postnatally. Since exposure of rat embryos younger than 11–12 days of age was not attempted, it is difficult to determine if the difference between chick and rat is due to species, dose or critical period differences.

Summary

Drugs were injected into fertilized chicken eggs at various stages of development. Depletion of catecholamines during embryogenesis resulted in significant increases in tyrosine hydroxylase activity long after hatching. Reserpine injection early in development also caused a significant increase in catecholamines at the same time enzyme activity was elevated. However, d,l-α-methyl-p-tyrosine failed to increase steady-state levels of catecholamines postnatally. Behavioral analysis postnatally showed opposite effects of treatment with the two agents, event though enzyme activity was elevated in both groups. Comparisons with actions of methylmercury in this species and reserpine given prenatally to rats are made.

References

1 *Bakir, F.; Damluji, S.F.; Amin-Zaki, L.; Murtadha, M.; Khalidi, A.; Al-Rawi, S.; Tikriti, S.; Dhahir, H.I.; Clarkson, T.W.; Smith, J.C., and Doherty, R.A.:* Methylmercury poisoning in Iraq. Science *230–241 (1973).*

2   Bartolomé, J.; Seidler, F.J.; Anderson, T.R., and Slotkin, T.A.: Effects of prenatal reserpine administration on development of the rat adrenal medulla and central nervous system. J. Pharmac. exp. Ther. *197:* 293–302 (1976).
3   Beani, L.; Ledda, F.; Bianchi, C., and Baldi, J.: Reversal by 3,4-dihydroxyphenylanine of reserpine-induced regional changes in acetylcholine content in guinea pig brain. Biochem. Pharmac. *15:* 779–784 (1966).
4   Bell, R.W.; Nitchke, W.; Bell, N.J., and Zachman, T.A.: Early experience, ultrasonic vocalizations and maternal responsiveness in rats. Devl Psychobiol. *7:* 235–242 (1974).
5   Bignami, G.: Behavioral pharmacology and toxicology. A. Rev. Pharmac. Toxic. *16:* 329–366 (1976).
6   Brown, P.L. and Jenkins, M.H.: Autoshaping of the pigeon's key peck. J. exp. Analysis Behav. *11:* 1–8 (1968).
7   Coyle, I.; Wayner, M.J., and Singer, G.: Behavioral teratogenesis: a critical evaluation. Pharmacol. Biochem. Behav. *4:* 191–200 (1976).
8   Coyle, J.T. and Axelrod, J.: Tyrosine hydroxylase in rat brain: developmental characteristics. J. Neurochem. *19:* 1117–1123 (1972).
9   Eiduson, S.: 5-Hydroxytryptamine in the developing chick brain: its normal and altered development and possible control by end-product repression. J. Neurochem. *13:* 923 (1966).
10 Hughes, J.A.; Rosenthal, E., and Sparber, S.B.: Time dependent effects produced in chicks after prenatal injection of methylmercury. Pharmacol. Biochem. Behav. *4:* 507–513 (1976).
11 Lydiard, R.B.; Fossom, L.H., and Sparber, S.B.: Postnatal elevation of brain tyrosine hydroxylase activity, without concurrent increases in steady-state catecholamine levels, resulting from dl-α-methylparatyrosine administration during embryonic development. J. Pharmac. exp. Ther. *194:* 27–36 (1975).
12 Lydiard, R.B. and Sparber, S.B.: Evidence for a critical period for postnatal elevation of brain tyrosine hydroxylase activity resulting from reserpine administration during embryonic development. J. Pharmac. exp. Ther. *189:* 370–379 (1974).
13 Lydiard, R.B. and Sparber, S.B.: Postnatal behavioral alterations resulting from prenatal administration of dl-alphamethylparatyrosine. Devl. Psychobiol. *10:* 305–314 (1977).
14 Murakami, U.: The effect of organic mercury on intrauterine life; in *Klingberg, Abramovici and Chemke* Drugs and fetal development, pp. 301–336 (Plenum Press, New York 1972).
15 Neims, A.H.; Warner, M.; Loughnan, P.M., and Arauda, J.V.: Developmental aspects of the heptatic cytochrome P450 monooxygenase system. A. Rev. Pharmac. Toxic. *16:* 427–445 (1976).
16 Newton, G. and Levine, S.: Early experience and behavior: the psychobiology of development (Thomas, Springfield 1968).
17 Paoletti, R. and Davison, A.N.: Chemistry and brain development. Adv. exp. Biol. Med., vol. 13 (Plenum Press, New York 1971).
18 Peters, J.; Vonderahe, A., and Powers, T.: The functional chronology in developing chick nervous system. J. exp. Zool. *133:* 505–518 (1956).
19 Seiden, L.S.; MacPhail, R.C., and Oglesby, M.W.: Catecholamines and drug-behavior interactions. Fed. Proc. Fed. Am. Socs exp. Biol. *34:* 1823–1831 (1975).
20 Sparber, S.B.: Effects of drugs on the biochemical and behavioral responses of developing organisms. Fed. Proc. Fed. Am. Socs exp. Biol. *31:* 74–80 (1972).
21 Sparber, S.B.: Neurochemical changes associated with schedule-controlled behavior. Fed. Proc. Fed. Am. Socs exp. Biol. *34:* 1802–1812 (1975).

22 *Sparber, S.B. and Shideman, F.E.:* Prenatal administration of reserpine: effect upon hatching, behavior and brain stem catecholamines of the young chick. Devl. Psychobiol. *1:* 236–244 (1968).
23 *Sparber, S.B. and Shideman, F.E.:* Detour learning in the chick: effect of reserpine administered during embryonic development. Devl Psychobiol. *2:* 56–59 (1969).
24 *Sparber, S.B. and Shideman, F.E.:* Estimation of catecholamines in the brains of embryonic and newly hatched chickens and the effects of reserpine. Devl Psychobiol. *2:* 115–119 (1969).
25 *Sparber, S.B. and Shideman, F.E.:* Elevated catecholamines in thirty-day-old chicken brain after depletion during development. Devl Psychobiol. *3:* 123–129 (1970).
26 *Spyker, J.M.; Sparber, S.B., and Goldberg, A.M.:* Subtle consequenes of methylmercury exposure: behavioral deviations in offspring of treated mothers. Science *177:* 621–623 (1972).
27 *Thoenen, H.; Mueller, R.A., and Axelrod, J.:* Trans-synaptic induction of adrenal tyrosine hydroxylase. J. Pharmac. exp. Ther. *169:* 249–254 (1969).
28 *Vernadakis, A. and Weiner, N.:* Drugs and the developing brain. Adv. Behav. Biol., vol. 8 (Plenum Press, New York 1974).
29 *Werboff, J. and Gottlieb, J.S.:* Drugs in pregnancy: behavioral teratology. Obstet. Gynaec. Survey *18:* 420–423 (1963).
30 *Young, R.D.:* Effect of prenatal drugs and neonatal stimulation on later behavior. J. comp. physiol. Psychol. *58:* 309–311 (1964).

S.B. Sparber, PhD, Department of Pharmacology, 105 Millard Hall, University of Minnesota, *Minneapolis, MN 55455* (USA)

Developmental Neurochemical and Behavioral Effects of Drugs[1]

Loy D. Lytle and Edwin Meyer, jr.

Massachusetts Institute of Technology, Department of Nutrition and Food Science, Cambridge, Mass.

Introduction

The results of research conducted previously in our laboratory (4, 12, 20, 22) and in others (5, 10, 11, 17, 19, 24) suggest that drug-induced changes in the behavior of nonprecocial mammals may depend on the age of the animal at the time of testing. For example, neonatal rats injected with the indirectly acting sympathomimetic drug amphetamine, show increases in locomotor activity from the time of birth (4, 12, 20, 24); in contrast, animals treated with drugs thought to alter brain cholinergic (3, 4, 10–12) or serotoninergic (24) neurotransmission do not show adult patterns of behavior until approximately the time of weaning.

The purpose of the present experiments was to determine whether the quality of amphetamine-induced hyperactivity was similar in animals of different ages, and to explore the possible extent to which differences in drug metabolism might play a role in the drug reactions of developing animals.

Material and Methods

Animals

Male albino rats, 10 or 100 days old at the time of testing, were bred and reared in the MIT vivarium from a parent stock of Charles River (Wilmington, Mass.) CD-derived Sprague-Dawley rats. At birth, litters were culled to 8 animals and were otherwise left undisturbed until the time of the experiment. Older rats were weaned at 21 days of age, and housed in groups of 6–8. All animals had *ad libitum* access to food (Purina Lab chow) and water, and were exposed to a constant room temperature (20–22 °C) and a standard 12:12 h light-dark cycle (Vita-Lite, DuroTest Corp., North Bergen, N.J.; lights on at 08.00 h).

[1] Supported in part by a grant from the National Institute of Mental Health (MH-25075), and by an Alfred P. Sloan Foundation Fellowship in Neurosciences to L.D.L.

Procedure

In the initial experiments, 10- or 100-day-old rats were habituated to photocell cages, scaled to the size of the animal (23), for 30 min. Following this adaptation period, different groups of animals (n = 8) were injected intraperitoneally with the 0.9% saline vehicle (1 ml/kg), or with 0.5, 1.0, or 2.0 mg/kg of d-amphetamine sulfate (salt weight; Sigma Chemical Co., St. Louis, Mo.). Locomotor activity was recorded automatically at hourly intervals for up to 10 h postinjection.

In the next experiment we determined whether the time course for the accumulation of amphetamine in the brains of 10- or 100-day-old animals differed. To accomplish this goal, animals were injected with 1.0 mg/kg (53 μCi/kg) of [^{3}H]-d-amphetamine sulfate, and were killed 5, 30, 60, 120, 240, or 480 min later. Brains were rapidly removed, frozen, and assayed for concentrations of amphetamine using a modification of the method of *Axelrod* (1).

In a final experiment, we determined the extent to which the metabolism of amphetamine might play a role in the drug's behavioral effects in young or adult animals. 10- or 100-day-old rats were habituated to the photocell cages for 30 min and then divided randomly into two groups. One group of animals was injected with 50 mg/kg of the hepatic microsomal enzyme inhibitor, diethylaminoethyl diphenylpropylacetate hydrochloride (SKF-525A) (6); the other group received an intraperitoneal injection of the 0.9% saline vehicle (1 ml/kg). 40 min later, half of each of these groups of animals received 1.0 mg/kg of d-amphetamine sulfate and the other half were administered the 0.9% saline vehicle. Locomotor activity was then recorded at hourly intervals for up to 10 h after the second injection.

Results

Amphetamine significantly increased the locomotor activity of neonatal or adult male rats in a dose-related fashion (fig. 1); however, there were clear age-related differences in response to the drug. Adult male rats injected with various doses of amphetamine showed peak increases in activity within 1–2 h postinjection, and these drug-induced increases in locomotor activity lasted for as long as 3 h following the highest dose of amphetamine (2.0 mg/kg) (fig. 1; bottom panel). In contrast, the peak increases in the drug-induced activity of the 10-day-old rats were less pronounced than those of the adult animals, and occurred slightly later (at 3–4 h postinjection). The duration of amphetamine's stimulant action was also greatly protracted in the younger animals at all doses tested; the activity levels of animals injected with 2.0 mg/kg of the drug were significantly elevated over vehicle baseline values for as long as 9 h post treatment (fig. 1; top panel).

The accumulation of amphetamine in the brains of neonatal or adult rats also showed age-related differences that were correlated temporally with both the peak behavioral effect of the drug, as well as with the duration of amphetamine-induced hyperactivity (fig. 2). Brain amphetamine concentrations were approximately equal in neonatal or adult male rats 30, 60, and 120 min postinjection; however, by 240 and 480 min following a single injection of the drug,

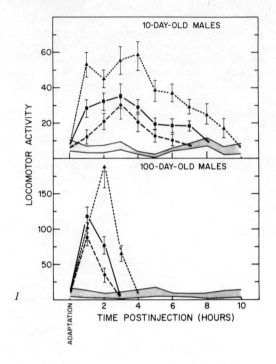

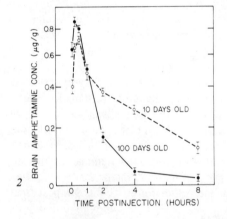

Fig. 1. Changes in locomotor activity in 10-day-old or 100-day-old male rats following an injection of the 0.9% saline vehicle (stippled area), or after 0.5 (●), 1.0 (■), or 2.0 (▲) mg/kg of *d*-amphetamine sulfate (salt weight). All values are the means ± SEM of at least 8 animals.

Fig. 2. Concentrations of amphetamine in the brains of 10- or 100-day-old male rats at various times after an injection of 1.0 mg/kg (53 µCi/kg) of [^{3}H]-*d*-amphetamine sulfate. All values are the means ± SEM of at least 6 animals.

Developmental Effects of Drugs 213

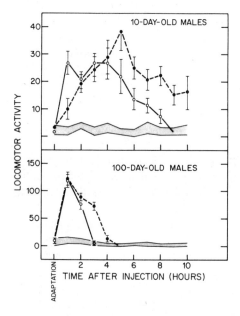

Fig. 3. Changes in locomotor activity in 10-day-old or 100-day-old male rats, pretreated with the 0.9% saline vehicle or 50 mg/kg of SKF-525A, followed 40 min later by an injection of saline or 1.0 mg/kg of d-amphetamine sulfate. There were no statistical differences in the locomotor activity patterns of animals receiving the saline−saline or SKF-525A−saline treatment combinations, and they have been combined for purposes of analysis (stippled area). The saline−amphetamine (o) and SKF-525A−amphetamine (•) treatment combination values are the means ± SEM of at least 8 animals.

brain amphetamine concentrations in neonatal animals were approximately 7 and 6 times greater, respectively, than in the 100-day-old adults (fig. 2). Hence, the half-life of amphetamine in the brains of rats appears to be age-related: whereas the drug half-life in the brains of adult male rats was approximately 44 min, it was 180 min in the brains of the neonatal animals.

Finally, it can be seen that the rate of amphetamine metabolism in either neonatal or adult rats influences its behavioral duration of action. 10- or 100-day-old animals pretreated with SKF-525A show protracted temporal behavioral responses to the stimulant effects of amphetamine when compared to appropriate aged rats receiving the saline control pretreatment (fig. 3). Amphetamine's stimulant duration of action is extended from 90 min in animals receiving the saline pretreatment to 150 min in rats pretreated with SKF-525A; similarly, the hepatic microsomal enzyme inhibitor extends the stimulant effects of amphetamine from 7 to over 9 h in the 10-day-old animals (fig. 3).

Discussion

In summary, then, the results of our experiments show that: (1) the stimulant effects of amphetamine are age-related, inasmuch as neonatal male rats show a later peak response, and longer duration of hyperactivity in response to amphetamine than do adult male rats; (2) the age-related differences in response to the behavioral effects of amphetamine are correlated with temporal differences in the accumulation of the drug in the brains of young versus adult animals; and (3) the relative rate at which amphetamine is metabolized in neonatal or adult rats determines, at least in part, its duration of stimulant action. Pretreatment with SKF-525A, a drug that inhibits the metabolism of amphetamine by blocking its *para*-hydroxylation, protracts the amphetamine-induced behavioral stimulation in both neonatal and adult rats.

Injections of low to moderate doses of amphetamine in adult animals result in a complicated array of physiological and behavioral changes, including psychomotor stimulation, a loss of appetite (anorexia), and changes in body temperature (8). Interestingly, the drug is a close structural analogue of the catecholamine neurotransmitters, and appears to produce most of its physiological and behavioral effects by increasing catecholaminergic neurotransmission among brain cells (8).

Not all of the physiological and behavioral effects of amphetamine are present in immature animals. Although the locomotor stimulant effects of the drug can be detected at birth, or shortly thereafter (4, 21), the anorexic and some of the thermogenic effects of the drug are not clearly evident in rats until the time of weaning (20–22). A wide variety of evidence indicates that catecholamine-containing neurons in altricial animals reach functional maturity relatively late in postnatal development, and the lack of mature physiological and behavioral responses to amphetamine in nonprecocial newborn animals may result from the relative immaturity of the neurotransmitter systems that mediate the action of the drug (20).

It is well known that immature animals also lack fully developed mechanisms for metabolizing, and thereby inactivating, a wide variety of drugs. For example, newborn rats metabolize the anesthetic drug pentobarbital more slowly than do adult animals, and this age-related difference in drug metabolism is associated with a longer-lasting drug-induced behavioral stupor in young animals (16). Amphetamine is normally inactivated in adult mammals primarily by liver microsomal enzymes which either deaminate (2, 9) or *para*-hydroxylate (2, 9) the parent compound. In the rat, amphetamine is *para*-hydroxylated in the liver to form p-hydroxyamphetamine; this metabolite is then conjugated, excreted, or further hydroxylated by the enzyme dopamine-β-hydroxylase to form p-hydroxynorephedrine (15). The rate at which amphetamine is inactivated influences its physiological and behavioral potency in adult animals (6, 7, 13, 15, 18). It is possible, then, that the protracted temporal responses of young rats to the stimulant effects of amphetamine seen in the present study may also be due to age-related differences in the rates at which amphetamine is metabolized and excreted (14). This may make more of the drug available to brain catecholaminergic synapses for longer periods of time in immature animals.

Summary

Amphetamine increases the locomotor activity of 10- or 100-day-old rats in a dose-related fashion. However, the duration of drug-induced psychomotor stimulation is greatly protracted in the younger animals. These age-related differences in response to the stimulant effects of amphetamine may result from the fact that young rats metabolize the drug more slowly than do adult animals, thus increasing the relative amount of the drug at active brain target sites.

References

1. *Axelrod, J.:* Studies on sympathomimetic amines. II. The biotransformation and physiological disposition of *d*-amphetamine, *d-p*-hydroxyamphetamine and *d*-methamphetamine. J. Pharmac. exp. Ther. *110:* 315–326 (1954).
2. *Axelrod, J.:* Amphetamine metabolism, physiological disposition and its effects on catecholamine storage; in *Costa and Garattini* International symposium on amphetamines and related compounds, pp. 207–216 (Raven Press, New York 1970).
3. *Blozovski, D. et Blozovski, M.:* Effets de l'atropine sur l'exploration, l'apprentissage et l'activité électrocorticale chez le rat au cours du développement. Psychopharmacologia *33:* 39–52 (1973).
4. *Campbell, B.A.; Lytle, L.D., and Fibiger, H.C.:* Ontogeny of adrenergic arousal and cholinergic inhibitory mechanisms in the rat. Science *166:* 635–637 (1969).
5. *Campbell, B.A. and Mabry, P.D.:* Ontogeny of behavioral arousal. A comparative study. J. comp. physiol. Psychol. *81:* 371–379 (1972).
6. *Clay, G.A.; Cho, A.K., and Roberfroid, M.:* Effect of diethylaminoethyl diphenylpropylacetate hydrochloride (SKF-525A) on the norepinephrine-depleting actions of *d*-amphetamine. Biochem. Pharmac. *20:* 1821–1831 (1971).
7. *Consolo, S.; Dolfini, E.; Garrattini, S., and Valzelli, L.:* Desipramine and amphetamine metabolism. J. Pharm. Pharmac. *19:* 253–256 (1967).
8. *Costa, E. and Garrattini, S.:* International symposium on amphetamines and related compounds (Raven Press, New York 1970).
9. *Dring, L.G.; Smith, R.L., and Williams, R.T.:* The metabolic fate of amphetamine in man and other species. Biochem. J. *116:* 425–435 (1970).
10. *Egger, G.J.; Livesey, P.J., and Dawson, R.G.:* Ontogenetic aspects of central cholinergic involvement in spontaneous alternation behavior. Devl Psychobiol. *6:* 289–299 (1973).
11. *Feigley, D.A.:* Effects of scopolamine on activity and passive avoidance learning in rats of different ages. J. comp. physiol. Psychol. *87:* 26–36 (1974).
12. *Fibiger, H.C.; Lytle, L.D., and Campbell, B.A.:* Cholinergic modulation of adrenergic arousal in the developing rat. J. comp. physiol. Psychol. *72:* 384–389 (1970).

13   *Freeman, J.J. and Sulser, F.:* Iprindole-amphetamine interactions in the rat. The role of aromatic hydroxylation of amphetamine in its mode of action. J. Pharmac. exp. Ther. *183:* 307–315 (1972).
14   *Groppetti, A. and Costa, E.:* Factors affecting the rate of disappearance of amphetamine in rats. Int. J. Neuropharmacol. *8:* 209–215 (1969).
15   *Groppetti, A. and Costa, E.:* Tissue concentrations of p-hydroxynorephedrine in rats injected with d-amphetamine. Effect of pretreatment with desipramine. Life Sci. *8:* 653–665 (1969).
16   *Kato, R.:* Sex-related differences in drug metabolism. Drug Metabolism Rev. *3:* 1–32 (1974).
17   *Kellogg, C. and Lundborg, P.:* Ontogenetic variations in responses to l-dopa and monoamine receptor-stimulating agents. Psychopharmacologia *23:* 187–200 (1972).
18   *Jonsson, J. and Lewander, T.:* Effects of diethyldithiocarbamate and ethanol on the *in vivo* metabolism and pharmacokinetics of amphetamine in the rat. J. Pharm. Pharmac. *25:* 589–591 (1973).
19   *Lal, S. and Sourkes, T.L.:* Ontogeny of stereotyped behaviour induced by apomorphine and amphetamine in the rat. Archs int. Pharmacodyn. Thér. *202:* 171–182 (1973).
20   *Lytle, L.D. and Keil, F.C.:* Brain and peripheral monoamines. Possible role in the ontogenesis of normal and drug-induced responses in the immature mammal; in *Fuxe, Olson and Zotterman* Dynamics of degeneration and growth in neurons, pp. 575–591 (Pergamon Press, New York 1974).
21   *Lytle, L.D.; McGuire, R.A.; Pettibone, D.; Courtright, W., and Schwartz, J.:* Developmental effects of 6-hydroxydopamine on temperature regulation in the rat; in *Jonsson, Malmfors and Sachs* Chemical tools in catecholamine research, vol. 1, pp. 189–196 (North-Holland, Amsterdam 1975).
22   *Lytle, L.D.; Moorcroft, W.H., and Campbell, B.A.:* Ontogeny of amphetamine anorexia and insulin hyperphagia in the rat. J. comp. physiol. Psychol. *77:* 388–393 (1971).
23   *Moorcroft, W.H.; Lytle, L.D., and Campbell, B.A.:* The ontogeny of hunger-induced arousal in the rat. J. comp. physiol. Psychol. *75:* 59–67 (1971).
24   *Mabry, P.D. and Campbell, B.A.:* Ontogeny of serotonergic inhibition of behavioral arousal in the rat. J. comp. physiol. Psychol. *86:* 193–201 (1974).

L.D. Lytle, PhD, University of California, Santa Barbara, Department of Psychology, *Santa Barbara, CA 93106* (USA)

Neurotransmitter Interactions and Early Convulsive Activity

A Developmental Model of Behavioral Responsivity[1]

C. Kellogg

Department of Psychology, University of Rochester, Rochester, N.Y.

Audiogenic (AG) seizures in inbred strains of mice have been selected to study neurochemical events contributing to a specific behavior. The seizure response to a high intensity, high frequency sound stimulus is a highly stereotyped, predictable response pattern. Following the onset of the stimulus, the animals demonstrate a sequential pattern of wild running, clonic seizures, tonic seizures, followed by death or recovery. The manifestation of AG seizures in genetically sensitive strains of mice follows age-specific response patterns. Mice begin to show some seizure responses to sound around 14 days postnatal age, and demonstrate maximal clonic-tonic seizures with a high lethality rate at 21 days. Dissipation of the behavior follows, and at 42 days, animals not previously tested demonstrate no severe seizure responses. AG seizures, therefore, provide a model in which to analyze the development of specific neurochemical processes in relationship to development of normal function or abnormal behavior.

The role of central neurotransmitters in the regulation of AG seizure activity has been extensively investigated pharmacologically (for review, see 10). A series of pharmacological experiments was undertaken in this laboratory to determine more precisely the contribution various central neurotransmitters make towards regulation of seizure susceptibility in the DBA/2J mice at their age of maximal sensitivity (10).

With respect to the catecholamines, the results indicated that not only was L-dopa (L-3,4-dihydroxyphenylalanine, the immediate precursor of noradrenaline, NA, and dopamine, DA) effective in protecting animals against seizures, but the NA agonist clonidine (2) was also effective, affording full protection at doses as low as 0.2 mg/kg. Apomorphine, a DA agonist (1), did not afford full protection against seizures even at doses as high as 5 mg/kg. However, apomor-

---

[1] This work was supported by Grant No. NS-10777 from the National Institute of Public Health, US Public Health Service.

phine did reduce the incidence of lethality, completely at the high dose and to 57% at 1 mg/kg. These results suggest that the protection against clonic-tonic seizures afforded by L-dopa is probably mediated via NA- rather than DA-containing neurons, although a role for DA-containing neurons in recovery from a seizure should not be dismissed.

Considering the indoleamine, 5-hydroxytryptamine (5-HT), protection against seizures can be afforded by the use of the 5-HT precursor 5-hydroxytryptophan (5-HTP) in the presence of pargyline, an inhibitor of monoamine oxidase. Because of possible interfering actions on catecholamine neurons by this drug combination, another method of manipulating 5-HT neurons was selected. Fluoxetine hydrochloride, an inhibitor of 5-HT uptake mechanisms (14), markedly reduced the incidence of lethality and prevented tonic seizures by interfering with tonic extension of the hind limbs. The combination of fluoxetine hydrochloride with 5-HTP further reduced the seizure index. However, even a high dose of 5-HTP (75 mg/kg) did not completely abolish the response to the sound stimulus. Such observations suggest that regulation of the AG seizure response by 5-HT neurons may be related to the motor aspects of the seizure.

Of the accepted central neurotransmitters, acetylcholine (ACh) has received little attention with respect to seizure protection. In our studies, the muscarinic agonist oxotremorine provides marked protection against seizures, even when the hypothermic effect of the drug is prevented by maintaining the animals in a warm environment. The central acting muscarinic antagonist scopolamine prevents the protection afforded by oxotremorine and by itself has little effect on seizure activity.

Hence, our pharmacologic studies have demonstrated that sound-induced clonic-tonic seizures and subsequent death can be prevented by drugs which alter neurotransmission in several different transmitter-containing neurons. The neuronal mechanisms regulating AG seizure susceptibility and severity in mice are thus quite complex, undoubtedly involving the interaction of several transmitter-containing neurons.

Based upon data gathered in our laboratory and upon the seizure literature (12), a schema illustrating possible interactions with respect to regulation of seizure activity and classical stress responses has been hypothesized (10) and is presented in figure 1. In this hypothesis, the sound stimulus is viewed as a stressor and the seizure represents a stress response. The NA neurons, when activated, would decrease the effect of a stressor and would receive the stress signal via cholinergic neurons which themselves could induce a stress response. In this way, NA neurons could monitor the stress level carried by ACh cells and respond by preventing excessive stress responses. Increased γ-aminobutyric acid (GABA) activation would inhibit the ACh-carried stress signal and decrease the need for the NA stress-dampening system.

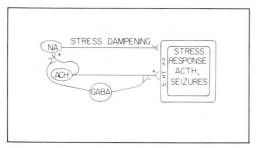

Fig. 1. Proposed scheme of transmitter interaction and regulation of stress responses.

These various transmitter-containing neurons achieve full functional maturity at different rates and for any one transmitter, the achievement of full maturity varies among brain regions. Normally then, the central responses to a stressor could be expected to vary developmentally and genetic factors could alter even this normal variation. Experiments to date have concentrated on analysis of the development of NA-containing neurons and muscarinic cholinergic receptors in the seizure-sensitive strain of mice, DBA/2J and seizure-resistant strain, C57BL/6.

A slower rate of maturation of NA levels has been observed in the seizure-sensitive DBA/2J mice, whereas there are no significant differences between strains in the maturation of DA levels (8). The *in vivo* catecholamine synthesis rate (7) was found to be significantly greater in the brains of the DBA/2J mice at 14 days. The rate decreased with age within this strain reaching a lower rate than in the C57BL/6 mice by 42 days. These results indicated developmental differences between strains in certain mechanisms of NA neurotransmission.

In vitro experiments on the development of the uptake of tritiated NA (^{3}H-NA) at $10^{-7} M$ demonstrated a significant decrease in the DBA/2J mice in the values for 20-min uptake in the telencephalon at their age of maximal seizure sensitivity (9). No age fluctuation was observed in the C57BL/6 mice. This marked decrease in 20-min uptake of ^{3}H-NA could represent deficiencies in releasable transmitter which may contribute to their marked seizure sensitivity at 21 days of age.

Recent studies have given support to an importance of NA in very young animals for protection against AG seizures. DBA/2J mice were given subcutaneous injections of the neurotoxin 6-hydroxydopamine (6-OHDA) at 50 mg/kg on days 5, 6 and 7 in order to deplete NA in terminal areas such as the cortex, hippocampus, and cerebellum (13). The animals were tested for seizure activity at 14, 21 and 42 days of age. The 6-OHDA treatment increased the severity of the seizure at 14 days, inducing a high incidence of tonic seizures and death similar to the response normally observed at 21 days (fig. 2). However, the

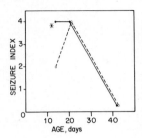

Fig. 2. Effect of neonatal 6-OHDA treatment on age-specific AG seizures. Data presented as modal values. ● = 6-OHDA, ○ = vehicle, * = $p < 0.05$.

dissipation of the behavior between 21 and 42 days was not altered by this treatment. The depletion of NA by 6-OHDA was monitored by measuring the 5-min uptake of ^{3}H-NA. Around a 40–50% reduction in uptake was observed in telencephalic and cerebellar minces from treated mice at all ages. Therefore, NA-containing neurons may be of marked importance in young animals for the regulation of stress responses, becoming of less importance as the animals mature. This suggestion is supported by observations that certain chemical alterations in NA neurons are still apparent at 42 days. Also, 6-OHDA treatment does not induce seizure sensitivity in the resistant C57BL/6 mice. Alterations in other transmitter-containing neurons, in addition to those changes noted in the NA neurons, must be contributing to the AG seizure susceptibility in the DBA/2J mice.

As indicated above the cholinergic muscarinic agonist oxotremorine affords protection against AG seizures in the DBA/2J mice at 21 days. We have also found this drug to prevent the acoustic induction of AG seizures in the C57BL/6 mice at 16 days of age. Whereas cholinergic inhibition of behavioral arousal has not been considered present until around 20 days of age in the rat (6) our studies suggest that muscarinic receptors involved in the regulation of AG seizures are functional at a much earlier age. Therefore, studies on the development of the cholinergic muscarinic receptor have been initiated (*Aronstam and Kellogg,* in preparation).

Development of the muscarinic receptor was approached by analyzing the binding of the muscarinic antagonist ^{3}H-3-quinuclidinyl benzilate (^{3}H-QNB) (15). In whole brain, there was an increase in the amount of ^{3}H-QNB bound with increasing age. The saturation curves were similar at all ages indicating no change in the affinity of the receptor with age for QNB. No strain differences were apparent in the whole brain studies. The binding of ^{3}H-QNB at 4×10^{-9} M was then analyzed in brain regions. Significant increases in total binding with age were observed in all regions except the midbrain-brain stem. In the hippocampus, the DBA/2J mice had significantly greater binding, and in the hypothalamus decreases in total binding with age were noted in the DBA/2J mice, whereas slight increases were noted in the C57BL/6 mice.

To further characterize the muscarinic receptors, inhibition of ^{3}H-QNB binding (5×10^{-11} M) by the agonist carbamylcholine was analyzed. The ability of carbamylcholine to compete with ^{3}H-QNB varied widely from area to area within both strains (fig. 3). Carbamylcholine affinity for muscarinic receptors was highest in the brain stem and hypothalamus. In the hippocampus of the DBA/2J mice, the inhibition of ^{3}H-QNB binding by carbamylcholine was substantially lower than in other regions, particularly at 14 days. With age, the percent inhibition increased within this strain, and by 42 days the amount of inhibition in this region was similar to that seen in the cortex, thalamus, and striatum. Within the C57BL/6 mice, this change with age was not observed.

Whereas the binding of tritiated antagonists to muscarinic receptors follows the law of mass action, indicating an interaction with a single population of binding sites, agonist binding does not follow the mass action isotherm (5). A model based upon the existence of two independent populations of agonist binding sites had been proposed (4). In the present study, the data presented in figure 3 was resolved into two components in order to determine the proportion of agonist binding sites in a high affinity versus a low affinity state.

In the brain stem and hypothalamus, a high proportion of agonist binding is in the high affinity form (fig. 4), with 99% of the sites in this form in the brain stem of the C57BL/6 mice at 42 days of age. In the DBA/2J mice, the proportion of sites in the high affinity state, while highest in the brain stem, is lower at all ages than was observed in the other strain. The most marked difference between strains, however, is the age-related shift in the percent of high affinity agonist sites noted in the hippocampus of the seizure-sensitive DBA/2J mice. At 14 days, there were no sites present in the high affinity state; by 21 days 11% were in this state and by 42 days 30% were in this state, a value comparable to that noted in the C57BL/6 strain at all ages.

The implication of different agonist binding sites is not understood. However, studies on the muscarinic receptor in smooth muscle suggest that the contractile response can be equated with occupancy of the lower affinity binding sites (3). The high affinity agonist binding sites are apparently inactive with respect to effecting a response. The possibility exists, therefore, that in the hippocampus of the seizure-sensitive DBA/2J mice, all agonist-binding sites are in the physiologically active form at 14 days, as they are all in the low affinity form. No until after the period of maximal seizure susceptibility in these mice does the proportion of high affinity sites reach a value comparable to that observed in the resistant strain. Also in the brain stem of the DBA/2J mice a greater portion of binding sites are in the low affinity state than in the resistant strain.

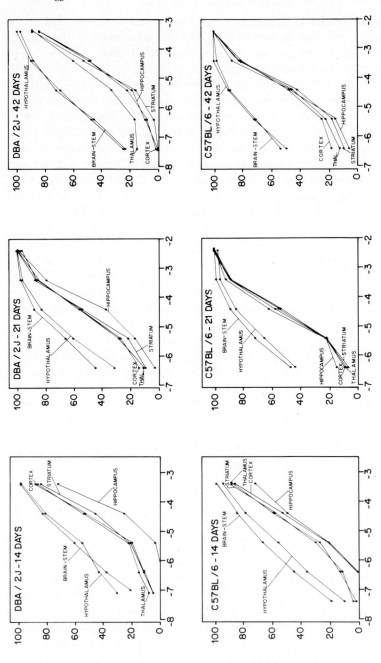

Fig. 3. Carbamylcholine inhibition of ^{3}H-QNB binding.

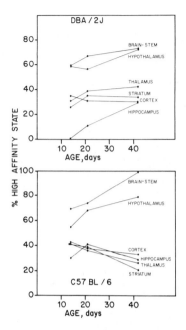

Fig. 4. Proportion of high affinity agonist binding sites. Estimated K_D was $2.65 \times 10^{-7}\ M$ as compared to a K_D of $1.04 \times 10^{-4}\ M$ for low affinity sites.

Our studies to date have indicated that central regulation of responses to a high intensity sound stimulus involves the complex interaction of several transmitter-containing neurons. The age-related aspects of the response may reflect the stage of development and interaction between the various neurons.

NA-containing neurons appear to be involved in determining the severity of the seizure in animals at 14 and 21 days, since depletion of NA in certain terminal areas by 6-OHDA increased the severity of the response in 14-day-old mice. A NA stress-dampening system may be most important for maintaining a proper level of neuronal activity during the period of rapid brain maturation.

However, the dissipation of the seizure response after 21 days of age does not appear related solely to correction of any deficiencies within NA neurons. Maturation of other systems is undoubtedly of importance. In our proposed scheme of transmitter interaction and regulation of stress responses, GABA neurons were proposed as inhibiting ACh neurons and thereby decreasing the need for a NA stress-dampening system. Hence maturation within GABA neurons may be critical for the reduction in AG seizures with age.

The results on the development of muscarinic receptors also demonstrated certain strain differences. The interpretation of these differences awaits further studies, however, the regional location of the differences in carbamylcholine inhibition of ^{3}H-QNB binding (the hippocampus and brain stem) suggests age-related changes in ACh neurons could be a determining factor in seizure severity. The hippocampus has classically been related to convulsive activity, and the brain stem contains several auditory relay nuclei. Additionally, evidence indicates that oxotremorine can elicit a prolonged activation of tyrosine hydroxylase in the locus ceruleus (11). Changing interaction etween NA and ACh with age may be of critical importance in determining age-specific AG seizure responses.

Summary

Audiogenic seizures in inbred strains of mice have been selected to study the development of neurotransmitters in relation to an age-specific behavioral response. Noradrenaline-containing neurons appear to be involved in determining the severity of the seizure in young animals. In particular depletion of noradrenaline in certain terminal areas by neonatal administration of 6-hydroxydopamine increases the severity of the response in 14-day-old mice but has no effect on the dissipation of the response after the usual peak response at 21 days. A noradrenaline-dampening system may be most important for maintaining a proper level of neuronal activity during the period of rapid brain maturation. Cholinergic agonists have been found to be protective against audiogenic seizures, suggesting involvement of central acetylcholine in the behavioral response. The development of the muscarinic receptor was studied and of particular interest was the observation of marked regional differences in the affinity of agonistic drugs for the muscarinic receptor. Also within the seizure-sensitive strain, there was a shift in agonist binding characteristics with age in the hippocampus. Not until after the age of peak sensitivity do the binding characteristics in this region become similar to those noted in the resistant strain at all ages. These studies have indicated that the central regulation of responses to a high intensity sound stimulus involves the complex interaction of several transmitter-containing neurons. The age-related aspects of the response may reflect the stage of development and interaction between the various neurons.

References

1 *Anden, N.-E.; Rubenson, A.; Fuxe, K., and Hökfelt, T.:* Evidence for dopamine receptor stimulation by apomorphine. J. Pharm. Pharmac. *19:* 627–729 (1967).
2 *Anden, N.-E.; Corrodi, H.; Fuxe, K.; Hökfelt, T.; Rydin, C., and Svensson, T.:* Evidence for a central noradrenaline stimulation by clonidine. Life Sci. *9:* 513–523 (1970).
3 *Birdshall, N.J.M.:* Muscarinic receptors. Biochemical studies. Biochem. Soc. Trans. *5:* 74–76 (1977).
4 *Birdshall, N.J.M. and Hulme, E.C.:* Biochemical studies on muscarinic acetylcholine receptors. J. Neurochem. *27:* 7–16 (1976).

5 *Birdshall, N.J.M.; Burgen, A.S.V.; Hiley, C.R., and Hulme, E.C.:* Binding of agonists and antagonists to muscarinic receptors. J. supramol. Struct. *4:* 376–371 (1976).
6 *Campbell, B.A.; Lytle, L.D., and Fibiger, H.C.:* Ontogeny of adrenergic arousal and cholinergic inhibitory mechanisms in the rat. Science *166:* 637–738 (1969).
7 *Carlsson, A.; Davis, J.N.; Kehr, W.; Lindqvist, M., and Atack, C.:* Simultaneous measurement of tyrosine and tryptophan hydroxylase activities in brain *in vivo* using an inhibitor of aromatic amino acid decarboxylase. Archs Pharmacol. *275:* 452–457 (1972).
8 *Kellogg, C.:* Audiogenic seizures. Relation to age and mechanisms of monoamine neurotransmission. Brain Res. *106:* 87–103 (1976).
9 *Kellogg, C. and Zuckerman, M.:* Age- and strain-related uptake of ^{3}H-noradrenaline into brain regions from audiogenic sensitive or resistant strains of mice. Neurosci. Abstr. *2:* 492 (1976).
10 *Kellogg, C. and Amaral, D.:* Neurotransmitter regulation of stress responses. Relationship to seizure induction; in *Butcher* Cholinergic-monoaminergic interactions in the brain (Academic Press, New York, in press).
11 *Lewander, T.; Joh, T., and Reis, D.:* Tyrosine hydroxylase. Delayed activation in central noradrenergic neurons and induction in adrenal medulla elicited by stimulation of central cholinergic receptors. J. Pharmac. exp. Ther. *200:* 523–534 (1977).
12 *Lovell, R.A.:* Some neurochemical aspects of convulsion; in *Lajtha* Handbook of neurochemistry, vol. 6, pp. 63–102 (Plenum Press, New York 1971).
13 *Sachs, C.; Pycock, C., and Jonsson, G.:* Altered development of central noradrenaline neurons during ontogeny by 6-hydroxydopamine. Med. Biol. *52:* 55 (1974).
14 *Wong, D.T.; Bymaster, F.P.; Horng, J.S., and Molloy, B.:* A new selective inhibitor for uptake of serotonin into synaptosomes of rat brain: 3-(*p*-trifluoromethylphenoxy)-N-methyl-3-phenylpropylamine. J. Pharmac. exp. Ther. *193:* 804–811 (1975).
15 *Yamamura, H.I. and Snyder, S.H.:* Muscarinic cholinergic receptor binding in the longitudinal muscle of the guinea pig ileum with (^{3}H)-quinuclidinyl benzilate. Mol. Pharmacol. *10:* 861–867 (1974).

C. Kellogg, PhD, Department of Psychology, University of Rochester, *Rochester, NY 14627* (USA)

Neurochemical Brain Changes Associated with Behavioural Disturbances after Early Treatment with Psychotropic Drugs[1]

P. Lundborg and J. Engel

Department of Pharmacology, University of Göteborg, Göteborg

Over the last years several studies have been published demonstrating behavioural changes in offspring after long-term prenatal and/or postnatal drug treatment. For example, modifications in the fetal brain that persist through adulthood and that are responsible for behavioural changes have been described after amphetamine (11, 16), imipramine (13), corticosterone (24), phenobarbital (23) and alcohol (7, 8).

From a clinical point of view it is of great interest to find out whether such behavioural changes can occur also after drug dosages low enough to be relevant for the clinical situation, and whether the behavioural changes can be related to specific CNS changes.

We have chosen to utilize the dopaminergic pathways of the brain as a model for such studies.

Although measurements of monoamine levels have been widely used to determine transmitter maturation, data on the mechanisms that control levels of monoamines in the brain yield more information. The study of developmental changes in the monoamine synthesizing enzymes and the monoamine uptake mechanisms (e.g. 12) as well as studies on the functional development of monoamine pathways (e.g. 19–21) seem to indicate that there is a time difference of development between noradrenaline (NA) and dopamine (DA) neurones. For example, in the rat the DA-containing neurones seem to gradually develop their functions during the first week of postnatal life, whereas the NA-containing neurones are more mature at birth. Based on these findings it was hypothesized that, in the rat and probably also the rabbit, the early postnatal period could be considered as a vulnerable period for the functional maturation of the central DA nervous system. Accordingly, the functional consequences of

---

[1] This research was sponsored by the Swedish Medical Research Council (No. 2464 and 4247), Swedish Board for Technical Development and Expressens Prenatalforskningsfond.

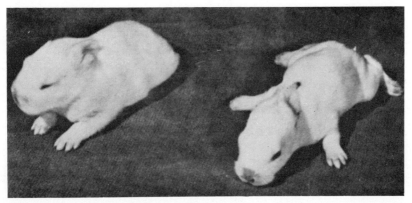

Fig. 1. 10-day-old rabbits nursed by mothers. Right, offspring of mother treated with haloperidol (1 mg/kg/day in drinking water from days 1–7 after parturition). Left, offspring of untreated mother. From *Kellogg et al.* (22).

giving drugs specifically affecting central DA mechanisms during this vulnerable period were to be tested.

In a first study young rabbits were exposed to haloperidol via chronic administration of the drug to their mothers during lactation (1 mg/kg/day for days 1–7 after birth). This treatment created a 'gait problem' in the offspring, with no behavioural effects on the mother (22) (fig. 1). At 14 days of age they were unable to right themselves, raise their heads or jump.

By 4 weeks of age most of the behavioural symptoms had abated but signs of dysfunction in the hind limbs remained. These results are interpreted as a consequence of a continuous blockade of central DA receptors in the brain during a period when central DA mechanisms are under maturation.

These studies were continued in rats using more specific behavioural tests (1–3) in combination with biochemical investigations (14, 15). Both penfluridol and pimozide are neuroleptic drugs known to block central catecholamine receptors, especially dopamine receptors (4, 17, 18). Either of these drugs was administered to nursing rat mothers and the result of the respective drug on the offspring 4 weeks after birth was examined by measuring the acquisition of a conditioned avoidance response. For details of the dose regimens, see legend to figure 2.

Also development of the animals were observed carefully as regards locomotion development, air-righting, startle response and eye-opening. No differences between the groups were found in the gross motor development after day 7. Neither were any statistically significant differences found in eye-opening and startle response. No differences were observed in litter mortality or in body

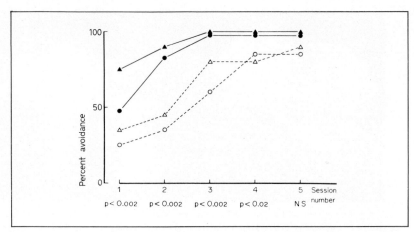

Fig. 2. Nursing rat mothers were given (a) pimozide (0.5 mg/kg i.p.) or 5.5% glucose (2 ml/kg i.p.) at days 1, 2, 3, 4, 5, 6 and 7 after delivery, or (b) penfluridol (1.0 mg/kg p.o.), or 5.5% glucose (5 ml/kg), at days 1, 3 and 5 after delivery. The offspring was trained to a CAR, 4 weeks after birth, in five consecutive daily sessions. Shown percent avoidance values are medians of 17 'pimozide-treated' infants. The p values refer to a statistical comparison between the 'pimozide'- and the 'glucose-treated' animals and were obtained by the Mann-Whitney U-test. ● = Pimozide controls, ○ = pimozide-treated, ▲ = penfluridol controls, △ = penfluridol-treated. From *Ahlenius et al.* (2).

weight. Furthermore, as assessed by gross observation, penfluridol-treated and glucose-treated mothers did not differ in their maternal behaviour as evidenced by nest-building, nursing and retrieval behaviour.

At the age of 28 days the male litter-mates were taken from their home cages. They were trained to avoid an electric shock (unconditioned stimulus, UCS) in a two-way shuttle box, with the sound of a house buzzer as a warning stimulus (conditioned stimulus, CS).

The male litter-mates were given five daily consecutive acquisition sessions, starting at the age of 28 days. Infants of mothers given glucose started with 50–70% CAR and reached 100% CAR from the time of the third training session. The infants of mothers given penfluridol showed an initial responding of only 35%, and after the fifth session they had reached the level of 90% avoidance responding. During every session a significant difference appeared between control and treated animals (fig. 2).

Hence treatment with any of the two DA-receptor-blocking agents penfluridol and pimozide appears to result in a persisting dysfunction in central catecholamine neurone systems which are involved in the acquisition of avoidance behaviour, this leading to the behavioural impairment observed. These findings support the hypothesis that the impairment observed is related to an

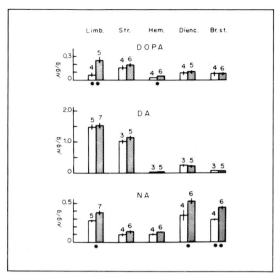

Fig. 3. Concentration of DOPA, DA and NA in limbic system (Limb.), striatum (Str.), hemispheres (Hem.), diencephalon (Dienc.), and brain stem (Br. st.) of 28-day-old infants of rat mothers treated with penfluridol (□; 1.0 mg/kg p.o.) or 5.5% glucose (5 ml/kg) (▨; controls). NSD-1015 (100 mg/kg i.p.) was given to all rats 30 min before death. From *Engel and Lundborg* (15).

inhibition of DA receptors during a period which could be considered vulnerable for the functional maturation of dopaminergic pathways in the brain. The impaired avoidance responding was in all probability not due to persistence of the drug in the young animal, since treatment of 28-day-old rats with penfluridol resulted in an impairment of the CAR that did not last more than 48 h (1).

Several investigations have been published during the last years emphasizing the importance of an undisturbed catecholamine neurotransmission for the maintenance of conditioned avoidance behaviour in adult rats and mice. Therefore large efforts have been made to find correlates between the behavioural changes in the 'penfluridol-treated' rats and biochemical changes in the brain.

For example, the *in vivo* rate of tyrosine hydroxylation, measured as the accumulation of 3,4-dihydroxyphenylalanine (DOPA) after inhibition of the aromatic aminoacid decarboxylase with NSD-1015 (10), was investigated.

Briefly, at an age of 28 days the male litter-mates, 'penfluridol-treated' as well as 'glucose-treated' were injected with NSD-1015, 100 mg/kg i.p. They were killed after 30 min. The brain was immediately dissected according to figure 3. The brain parts were analyzed for DOPA, DA and NA. A pronounced decrease in the accumulation of DOPA following decarboxylase inhibition was found in the limbic system (25% of controls) of young from the 'penfluridol-treated' rats (fig. 3). A slight decrease (about 60% of controls) was found also in the hemispheres.

No differences in the levels of DA were observed between young from 'penfluridol-treated' rats and controls in any of the brain regions examined (fig. 3). A statistically significant decrease of NA was, however, observed in the limbic system, diencephalon and brain stem of offspring from the 'penfluridol-treated' rats (fig. 3).

Hence we have shown that at the time of the impaired acquisition of a conditioned avoidance response, there are regional biochemical changes in the brains of the offspring of 'penfluridol-treated' rats. It is hard to draw any conclusions about the functional importance of changes in brain amine levels, like those observed for NA. Studies on the accumulation of DOPA after NSD-1015 is, however, a direct *in vivo* method for investigating the hydroxylation of tyrosine, the rate-limiting step for catecholamine synthesis (10), and a change in the DOPA accumulation as observed in the limbic system of 'penfluridol-treated' rats should be of pronounced functional importance. The nature of this biochemical brain damage can be further investigated by administration of α-methyl-*p*-tyrosine, an inhibitor of tyrosine hydroxylase (15). Animals were treated as described under figure 4.

A regression coefficient was calculated for the disappearance curves for DA and NA from all brain regions and tested for significance of linearity. In a comparison between 'penfluridol-treated' and 'glucose-treated' animals for each brain region the only difference found was a pronounced retardation ($p < 0.025$) in the rate of disappearance of DA from the limbic system of the 'penfluridol-treated' animals. It should be noted that in the DA-rich striatum no changes were observed.

The rate by which the monoamines disappear after complete inhibition of their biosynthesis is generally considered to reflect the impulse activity in the different monoamine neurones (5). Thus the findings in the present series of experiments may indicate that there is a reduced nerve impulse activity in the mesolimbic DA-neurones in the offspring of nursing mothers treated with penfluridol. It has repeatedly been shown in adult rats that inhibition of the impulse flow, pharmacologically or mechanically, of the central DA-neurones is followed by an increase in the DA synthesis, possibly mediated via autoreceptors on the DA nerve terminals (9). In the 'penfluridol-treated' animals we found, however, both a decreased impulse flow (measured as the disappearance of DA after inhibition of tyrosine hydroxylase) and a decreased DA synthesis (measured as the accumulation of DOPA after inhibition of aromatic amino acid decarboxylase). A possible explanation for this discrepancy may be disturbed feedback mechanism, caused by the treatment with a DA receptor antagonist early in development, i.e. during a vulnerable period for the functional maturation of the central DA systems.

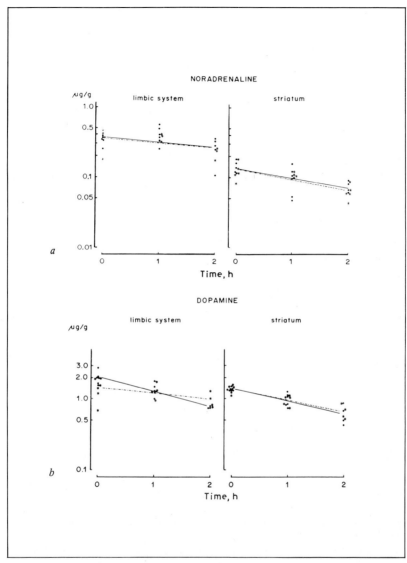

Fig. 4. Concentration of (a) noradrenaline ($\mu g/g$) and (b) dopamine ($\mu g/g$) in the limbic system and striatum following inhibition of tyrosine hydroxylase with α-methyltyrosine. Nursing rat mothers were given penfluridol (1 mg/kg p.o.) or 5.5% glucose (5 ml/kg) at day 1, 3 and 5 after delivery. At 28 days of age the male litter-mates were given α-methyltyrosine (250 mg/kg s.c.) or saline. 0 value represents saline-treated animals. Dashed lines represent calculated regression lines for offspring of 'penfluridol-treated' mothers and solid lines represent those of 'glucose-treated' mothers. Plotted are individual values. From *Engel and Lundborg* (16).

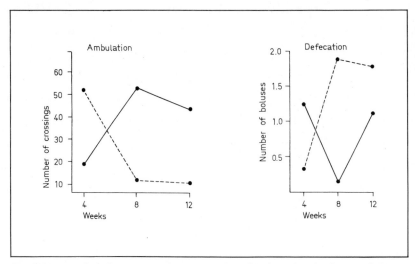

Fig. 5. Mean ambulation score and mean number of boluses in a 3-min open field test shown at 4, 8 and 12 weeks of age by offspring of nursing mothers treated with penfludilol (– – –) compared to controls (———). From *Ahlenius et al.* (3).

In addition to the test situations already described the open field behaviour of the 'penfluridol-treated' rats has been investigated (3). The animals were placed in a circular area (0.48 m²), surrounded by a 75 cm high wall and divided into 13 subareas of approximately equal size. The rats were observed at three 3-min sessions, each separated by an interval of approximately 30 min. The locomotor activity was recorded once a minute by counting the number of times an animal moved with all four legs into a new subarea. The number of boluses left in the area were counted after each test. The offspring were observed at the age of 4, 8 and 12 weeks.

It was found that at 4 weeks of age the 'penfluridol-treated' animals displayed a higher locomotor activity than the controls (fig. 5). At 8 and 12 weeks of age, when the activity of the control rats had increased markedly in comparison to the activity displayed at 4 weeks of age, the 'penfluridol-treated' rats showed a decrease to a level significantly lower than that of the controls. The number of boluses left in the area were counted after each test and showed a picture inverse to that of the ambulation score (fig. 5). Also, marked group differences were found in changes in locomotor activity occurring over the three sessions at the various age levels (fig. 6). At 4 weeks of age both the control and the 'penfluridol-treated' rats showed a systematic decrease in ambulation over the three sessions indicating intersession habituation. At 8 and 12 weeks, the control group showed a similar decrease in activity as at 4 weeks whereas the

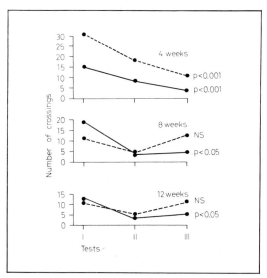

Fig. 6. Ambulation activity during the first minute in each of three consecutive 3-min open field tests in test performed during 1 day at 4, 8 and 12 weeks of age, displayed by experimental (– – –) and control rats (——). From *Ahlenius et al.* (3).

'penfluridol-treated' rats failed to show any sign of habituation and the activity of the 'penfluridol-treated' group was as high in the first session as in the last.

Low ambulation in an aversive open field situation combined with high defecation is traditionally thought to reflect increased emotionality or a lowered ability to adapt to novel stimuli (6). The high activity and low defecation scores shown by the 4-week-old 'penfluridol-treated' rats thus suggest an abnormally lowered emotional reactivity in these animals. The lowered ambulation, the increased defecation and the lowered ability to habituate indicate an abnormally heightened emotional reactivity with advancing age as a persistent consequence of the neonatal drug treatment. It should be mentioned that in the strains of rats used in these experiments, puberty starts around 60 days of age.

Summarizing these data from this laboratory, we have found the neonatal penfluridol treatment produces: (1) deficits in the acquisition of an active avoidance response; (2) an increased locomotor activity prepuberally followed by an abnormally decreased activity postpuberally, and (3) an impaired ability to habituate to novel stimuli. Associated with these behaviour changes we have found a decreased functional activity in the mesolimbic DA system at 4 weeks of age (prepuberal age). In support of the assumption that this behavioural syndrome is related to a dysfunction of the central DA system, we have found that the learning deficits could be counteracted by amphetamine (2), which acts by increasing the functional activity of the catecholamine neurones.

Summary

Neonatal rats have been treated with dopamine receptor blocking agents during the first postnatal week. This treatment resulted in deficits in the acquisition of an active avoidance response and an increased locomotor activity prepuberally followed by an abnormally decreased activity postpuberally. Associated with these behaviour changes we have found a decreased functional activity in the mesolimbic dopamine system at 4 weeks of age.

References

1. *Ahlenius, S.; Brown, R.; Engel, J., and Lundborg, P.:* Learning deficits in 4 weeks old offspring of the nursing mothers treated with the neuroleptic drug penfluridol. Arch. Pharmakol. *279:* 31–37 (1973).
2. *Ahlenius, S.; Engel, J., and Lundborg, P.:* Antagonism by d-amphetamine of learning deficits in rats induced by exposure to antipsychotic drugs during early postnatal life. Arch. Pharmakol. *288:* 185–193 (1975).
3. *Ahlenius, S.; Engel, J.; Hård, E.; Larsson, K.; Lundborg, P., and Sinnerstedt, P.:* Open field behaviour and gross motor development in offspring of nursing mothers given penfluridol. Pharmacol. Biochem. Behav. *6:* 343–347 (1977).
4. *Andén, N.-E.; Butcher, S.G.; Corrodi, H.; Fuxe, K., and Ungerstedt, U.:* Receptor activity and turnover of dopamine and noradrenaline after neuroleptics. Eur. J. Pharmacol. *11:* 303–314 (1970).
5. *Andén, N.-E.; Corrodi, H.; Fuxe, K., and Ungerstedt, U.:* Importance of nervous impulse flow for the neuroleptic induced increase in amine turnover in central dopamine neurons. Eur. J. Pharmacol. *15:* 193–199 (1971).
6. *Archer, J.:* Tests for emotionality in rats and mice. A review. Anim. Behav. *21:* 205–235 (1973).
7. *Bond, N.W. and Di Giusto, E.L.:* Effects of prenatal alcohol consumption on open-field behaviour and alcohol preference in rats. Psychopharmacology *46:* 163–165 (1976).
8. *Bond, N.W. and Di Giusto, E.L.:* Prenatal alcohol consumption and open-field behaviour in rats. Effects of age at time of testing. Psychopharmacology *52:* 311–312 (1977).
9. *Carlsson, A.:* Drugs which block the storage of 5-hydroxytryptamine and related amines; in *Eichler and Farah* Handbook of experimental pharmacology, vol. 19, pp. 529–592 (Springer, Berlin 1966).
10. *Carlsson, A.:* Dopaminergic autoreceptors; in *Almgren and Carlsson* Chemical tools in catecholamine research II, pp. 219–225 (North-Holland, Amsterdam 1975).
11. *Carlsson, A.; Davis, J.N.; Kehr, W.; Lindqvist, M., and Atack, C.V.:* Simultaneous measurements of tyrosine and tryptophan hydroxylase activities in brain *in vivo* using an inhibitor of the aromatic amino acid decarboxylase. Arch. Pharmakol. *275:* 153–168 (1972).
12. *Clark, C.; Gormen, D., and Vernadakis, A.:* Effects of prenatal administration of psychotropic drugs on behavior of developing rats. Devl Psychobiol. *3:* 225–235 (1970).
13. *Coyle, J.T.:* Development of the central catecholaminergic neurones in the rat; in *Usdin and Snyder* Frontiers in catecholamine research 1973, pp. 261–265 (Pergamon Press, Oxford 1973).
14. *Coyle, I.R. and Singer, G.:* The interactive effects of prenatal imipramine exposure and

postnatal rearing conditions on behaviour and histology. Psychopharmacology *44:* 253–256 (1975).
15 *Engel, J. and Lundborg, P.:* Regional changes in monoamine levels and the rate of tyrosine and tryptophan hydroxylation in 4 weeks old offspring of the nursing mothers treated with the neuroleptic drug penfluridol. Arch. Pharmakol. *282:* 327–334 (1974).
16 *Engel, J. and Lundborg, P.:* Reduced turnover in mesolimbic dopamine neurons in 4-week-old offspring of nursing mothers treated with penfluridol. Brain Res. *110:* 407–412 (1976).
17 *Hitzemann, B.A.; Hitzemann, R.J.; Brase, D.A., and Loh, H.H.:* Influence of prenatal d-amphetamine administration on development and behavior of rats. Life Sci. *18:* 605–612 (1976).
18 *Janssen, P.A.J.; Niemegers, C.J.E.; Schellekens, K.H.L.; Dresse, A.; Lenaerts, F.M.; Pinchard, A.; Schaper, W.K.A.; Van Neuten, J.M., and Verbruggen, F.J.:* Pimozide, a chemically novel, highly potent and orally long-acting neuroleptic drug. Arzneimittel-Forsch. *18:* 261–287 (1968).
19 *Janssen, P.A.J.; Niemegers, C.J.E.; Schellekens, K.H.L.; Lenaerts, F.M.; Verbruggen, F.J.; Van Neuten, J.M., and Schaper, W.K.A.:* The pharmacology of penfluridol (R 16341), a new potent and orally long-acting neuroleptic drug. Eur. J. Pharmacol. *11:* 139–154 (1970).
20 *Kellogg, C. and Lundborg, P.:* Ontogenic variations in response to L-DOPA and monoamine receptor-stimulating agents. Psychopharmacology *23:* 187–200 (1972).
21 *Kellogg, C. and Lundborg, P.:* Inhibition of catecholamine synthesis during ontogenic development. Brain Res. *61:* 321–329 (1973).
22 *Kellogg, C.; Lundborg, P., and Roos, B.E.:* Ontogenic changes in cerebral concentrations of homovanillic acid in response to haloperidol treatment. Brain Res. *40:* 469–475 (1972).
23 *Lundborg, P.:* Abnormal ontogeny in young rabbits after chronic administration of haloperidol to the nursing mother. Brain Res. *44:* 684–687 (1972).
24 *Middaugh, L.D.; Santos, C.A., and Zemp, J.W.:* Effects of phenobarbital given to pregnant mice on behavior of mature offspring. Devl Psychobiol. *8:* 305–313 (1975).
25 *Olton, D.S.; Johnsson, C.T., and Howard, E.:* Impairment of conditioned active avoidance in adult rats given corticosterone in infancy. Devl Psychobiol. *8:* 55–61 (1974).

Dr. *Per Lundborg,* Department of Pharmacology, University of Göteborg, Fack, *S–400 33 Göteborg 33* (Sweden)

Author Index

Arnold, E.B. 160

Black, I.B. 65
Blindermann, J.M. 83
Bondy, S.C. 116

Campochiaro, P. 134
Coughlin, M.D. 65
Coyle, J.T. 134

Deanin, G.G. 142
Di Renzo, M.F. 127
Duell, M.J. 152

Engel, J. 226

Fambrough, D.M. 31
Fernandez, H.L. 152
Festoff, B. 152
Filogamo, G. 1

Gardner, J.M. 31
Ghandour, M.S. 10
Giacobini, E. 41
Giacobini, G. 23
Goedert, M. 76

Gombos, G. 10
Gordon, M.W. 142

Hanson, R.K. 142
Harrington, M.E. 116
Hoffman, D.W. 160
Hösli, E. 108
Hösli, L. 108

Kellogg, C. 217
Koenig, H.L. 91
Koenig, J. 91
Krebs, H. 171

Lauder, J.M. 171
Levi, A. 142
London, E.D. 134
Lundborg, P. 226
Lytle, L.D. 210

Maitre, M. 83
Mandel, P. 83
Marchisio, P.C. 127
Meyer, E., jr. 210

Nidess, R. 116

Ossola, L. 83
Otten, U. 76

Peirone, S. 1
Pierce, T. 142
Purdy, J.L. 116

Reeber, A. 10

Sisto Daneo, L. 1
Sparber, S.B. 200
Suzuki, O. 100

Thoenen, H. 76
Tikkanen, I.T. 191
Timiras, P.S. 181
Tissari, A.H. 191

Vaccari, A. 181
Vernadakis, A. 160
Vincendon, G. 10

Yagi, K. 100

Zanetta, J.-P. 10